Revolutions in Twentieth-Century Physics

The conceptual changes wrought by modern physics are radical, important, and fascinating, yet they are only vaguely understood by people working outside the field. Exploring the four pillars of modern physics – relativity, quantum mechanics, elementary particles, and cosmology – this clear and lively account, that will interest anyone who has wondered what Einstein, Schrödinger, Feynman, Hubble, and others were really talking about.

The book discusses quarks and leptons, antiparticles and Feynman diagrams, curved spacetime, the Big Bang and the expanding Universe. Suitable for undergraduate students on nonscience majors as well as science majors, it uses problems and worked examples to help readers develop an understanding of what recent advances in physics actually mean.

DAVID J. GRIFFITHS is an Emeritus Professor of Physics at Reed College. He is the author of three highly regarded physics textbooks: *Introduction to Electrodynamics* (fourth edition, Pearson, 2012), *Introduction to Quantum Mechanics* (second edition, Pearson, 2004), and *Introduction to Elementary Particles* (second edition, Wiley-VCH, 2008).

Revolutions in Twentieth-Century Physics

DAVID J. GRIFFITHS
Reed College

CAMBRIDGE
UNIVERSITY PRESS

University Printing House, Cambridge CB2 8BS, United Kingdom

One Liberty Plaza, 20th Floor, New York, NY 10006, USA

477 Williamstown Road, Port Melbourne, VIC 3207, Australia

314-321, 3rd Floor, Plot 3, Splendor Forum, Jasola District Centre, New Delhi-110025, India

79 Anson Road, #06-04/06, Singapore 079906

Cambridge University Press is part of the University of Cambridge.

It furthers the University's mission by disseminating knowledge in the pursuit of education, learning and research at the highest international levels of excellence.

www.cambridge.org
Information on this title: www.cambridge.org/9781107602175

First published 2013
Reprinted 2015

A catalogue record for this publication is available from the British Library

Library of Congress Cataloging in Publication data
Griffiths, David J. (David Jeffery), 1942–
Revolutions in twentieth-century physics / David Griffiths, Reed College.
pages cm
Includes index.
ISBN 978-1-107-60217-5 (pbk.)
1. Quantum theory – Popular works.   2. General relativity (Physics) – Popular works.
I. Title.
QC174.12.G75154   2013
530 – dc23   2012030789

ISBN 978-1-107-60217-5 Paperback

Contents

Preface

The twentieth century saw four astonishing revolutions in physics: relativity, quantum mechanics, elementary particles, and cosmology. Each one radically changed our understanding of the Universe. There were also, of course, extraordinary breakthroughs in technology (electronics, lasers, computers) that had a much larger influence on our daily lives, but did not carry the same conceptual impact.

This is a book about those four revolutions. It is intended for anyone with a serious interest in the great ideas that have shaped modern physics: advanced high school students or freshman physics majors who would like a taste of what lies ahead; undergraduates who do not intend to major in the sciences but are curious to know about some of the most profound intellectual achievements of our time; general readers who have heard about quarks and quanta, Albert Einstein and the Big Bang, and would like to know what all the fuss is about.

I should tell you up-front what this book is *not*. It is not another breathless account of the fantastic speculations that seem to dominate much of contemporary theoretical physics – things you may have read about, or seen on NOVA. Apart from a few footnotes and an occasional parenthetical remark, there is nothing here about superstrings or extra dimensions or multiple universes. We're dealing with well-established, robustly confirmed "facts." In a way, modern physics has been a victim of its own success. The revolutions described in this book account so perfectly for everything that is known about our world, that anyone hoping to come up with the next "great idea" is forced to rely more on imagination than on observation. There is nothing wrong with this – inspired conjecture occupies an honored place at the very pinnacle of scientific discovery. Pauli "predicted" the neutrino, Dirac the antiparticle, Yukawa the meson, and Gell-Mann the omega-minus and the quark, using nothing but pencil and paper, well before these particles were detected in the laboratory.

But I do think it is important to distinguish between "knowledge" and "speculation." Some authors leave the impression that you could buy a six-pack of black holes and a bag of Higgs bosons at any convenience store. No doubt many of the current conjectures will turn out to be true, but that's not what this book is about. It's about the rock solid foundations on which *any* future developments must inevitably rest. They are already astounding enough to make a great story.

This book is not for casual bed-time reading, and it is not for everyone. You need to know some mathematics – arithmetic for sure, and a little algebra in places. There is no honest way to explain this material without it. But if you are intimidated by the sight of an equation, please don't give up too quickly. Physics tends to seem extremely difficult when you first encounter it, but quite simple once you understand it. Understanding comes with familiarity and practice. There are quite a few problems sprinkled through the text, and I urge you to work them all. Students often tell me, "I understand the *concepts*; I just can't do the problems." They are fooling themselves. The only sure test of whether you understand the concepts is precisely your ability to work the problems. But it *does* take time and practice. There are no short cuts.

Modern physics is built on a venerable foundation going back to Galileo and Newton, so I begin with a survey of the essential ideas, laws, and terminology we inherit from the classical era. I don't pretend this is anywhere near complete – I will concentrate on those pieces of the subject that are essential to the story that follows. We need to know about mass and velocity, force and energy, momentum and wavelength, gravity and electric charge. These things are the focus of Chapter 1. The subsequent chapters treat relativity, quantum mechanics, elementary particles, and cosmology, in that order. In principle the sequence of these topics is interchangeable, but I think this roughly chronological ordering makes the best sense. However, if you are only interested in one or two of them, you should be able to read the chapters independently.

I thank the students in my Master of Arts in Liberal Studies (MALS) course, for whom much of this was written, and especially Trina Marmarelli, who read the first draft with great care and fixed many awkward passages.

This book draws on material from my three advanced undergraduate textbooks: *Introduction to Electrodynamics*, 3rd ed., (Pearson, 1999. Printed and electronically reproduced by permission of Pearson Education, Inc., Upper Saddle River, New Jersey); *Introduction to Quantum Mechanics*, 2nd ed., (Pearson, 2005. Printed and electronically reproduced by permission of Pearson Education, Inc., Upper Saddle River, New Jersey); *Introduction to Elementary Particles*, 2nd ed., (Wiley-VCH, 2008. Copyright Wiley-VCH Verlag GmbH & Co. KgaA). Material from these textbooks is reproduced with the kind permission of Pearson and Wiley-VCH.

1

Introduction: classical foundations

1.1 Preliminaries

1.1.1 Units

If I ask you, "How long is an Olympic swimming pool?", and you answer "164," you're correct, in a way, but your response is worthless to me, because I don't know whether you are talking about *feet*, or *meters*, or *yards*, or *light-years*. Most physical quantities carry **dimensions** (length, time, mass, etc.), and you must indicate the **units** you are using. In this book we will for the most part use the metric system: meters (m), seconds (s), kilograms (kg), and so on. This is arbitrary, of course, and you can use inches, hours, and pounds, if you prefer – as long as you are careful to include the units whenever you specify a physical quantity. (You *should* have said "164 feet.") In fact, it's a good practice to carry the units along at each step in a problem: if you're calculating a distance, and it comes out in seconds, you *know* you've made a mistake, and there is no point in continuing; go back and find the error before moving on.

Because not everybody uses the same units, it is important to be able to **convert** from one system to another. This is easy, if you systematically replace each unit with its equivalent in the new system. For example, 1 inch (in) is 2.54 centimeters (cm), 1 m = 100 cm, and there are 12 inches in a foot (ft), so the length of that pool is

$$164\,\text{ft} = (164) \times 12\,\text{in} = (164)(12) \times 2.54\,\text{cm}$$
$$= (164)(12)(2.54) \times \frac{1}{100}\,\text{m} = 50\,\text{m}.$$

Example 1. How many square feet are there in a square yard?

Solution: A yard (yd) is 3 feet, so

$$1\,\text{yd}^2 = (1) \times (3\,\text{ft})^2 = 9\,\text{ft}^2.$$

1

So there are 9 square feet in a square yard. (If this surprises you, draw a 3 × 3 checkerboard and count the squares.)

Problem 1. What is 60 mph (miles per hour) in meters per second? (A mile is 5280 ft.)

Problem 2. A milliliter (0.001 l) is a cubic centimeter. How many liters are there in a cubic meter?

1.1.2 Scientific notation

We're going to encounter some huge numbers, and some very tiny numbers (the age of the Universe is 432 000 000 000 000 000 s; the radius of a hydrogen atom is 0.000 000 000 0529 m). It's hard to make sense of numbers like that – you have to squint even to *count* all those zeros. Much better is the "power-of-ten" notation. Notice that

$$10^1 = 10,$$
$$10^2 = 10 \times 10 = 100,$$
$$10^3 = 10 \times 10 \times 10 = 1000;$$

evidently the power of ten counts the number of zeros to the right of the number 1. As you increase the power of ten by one, you are *multiplying* by 10. Going in the other direction (reducing the power of ten) you are *dividing* by 10:

$$10^0 = 10/10 = 1,$$
$$10^{-1} = 1/10 = 0.1,$$
$$10^{-2} = 1/10^2 = 1/100 = 0.01,$$

and so on; a *negative* power tells you how many places to the right of the decimal point the 1 lies. In this language I can write the age of the Universe as 4.32×10^{17} s (the decimal point is 17 places to the right of the 4) and the radius of hydrogen as 5.29×10^{-11} m (the decimal point is 11 places to the *left* of the 5).

To multiply numbers, in this notation, you multiply the numbers out front and add the exponents:

$$(2 \times 10^3) \times (3 \times 10^4) = (2 \times 3) \times 10^{(3+4)} = 6 \times 10^7;$$
$$(3.14 \times 10^{-2}) \times (6.47 \times 10^{-3}) = 20.3 \times 10^{-5} = 2.03 \times 10^{-4}.$$

(It is customary to leave just one digit to the left of the decimal point, and let the power of 10 soak up the rest.) To divide, you divide the numbers out front, and subtract the exponents. If you wonder where these rules come from – or you forget how to do it – just make up a simple example for yourself, and you'll quickly figure it out:

$$400 \div 20 = \frac{400}{20} = \frac{4 \times 10^2}{2 \times 10^1} = \frac{4}{2} \times 10^{(2-1)} = 2 \times 10 = 20. \checkmark$$

By the way, *adding* (or subtracting) two numbers is awkward, in this notation. You first need to express both of them as the *same* power of 10:

$$419 + 23 = (4.19 \times 10^2) + (0.23 \times 10^2) = (4.19 + 0.23) \times 10^2$$
$$= 4.42 \times 10^2 = 442. \checkmark$$

This all takes some getting used to, so if it is new to you, make sure you can solve the following problems. Once you are confident you understand how it works, get a calculator, and let it take care of the details!

Problem 3. (a) How many seconds are there in a year? (b) What is the age of the Universe, in years?

Problem 4. A wheat field is 3 miles long and 2 miles wide. How many square inches is that?

Problem 5.

(a) $(3 \times 10^6) \times (12 \times 10^7) = ?$

(b) $\dfrac{(12 \times 10^{17})}{(4 \times 10^{13})} = ?$

(c) $(6.29 \times 10^4) + (7.1 \times 10^3) = ?$

1.1.3 Significant digits

Suppose I go to the blackboard, and, swinging my arm around, draw a big circle. *Question:* What's the circumference of that circle? Well, $c = 2\pi r$, where r is the radius, and I estimate my arm's length at about 70 cm (0.7 m). Punching this into my calculator (which has stored the value of π), I get

$$c = 4.398\,229\,715 \text{ m}.$$

What do you make of that result? Do you really believe the 5 down there in the ninth decimal place? After all, the value for r I used was only a rough estimate. It could easily be that my arm's length is actually 74 cm (in which case $c = 4.649\,557\,127$ m), or perhaps 67 cm ($c = 4.209\,734\,156$ m). Clearly the leading 4 is right, and the next digit seems to lie somewhere between 2 and 6, but all the rest is meaningless junk. I *should* have recorded the answer as

$$c = 4.4 \, \text{m}.$$

Only those two digits bear any relation to the truth – we call them **significant digits**.

In an era when calculators and computers happily spit out 8 or 16 digits, it is tempting to list them all, even when a moment's reflection indicates that most of them are insignificant. It's not *wrong*, exactly, but it's grossly misleading, and it looks unprofessional. Don't do it. If you want to carry one or two extra digits, just to be on the safe side, I won't quarrel with you, but not more than that. The reader naturally assumes that whatever numbers you list are valid. More precisely, if you write 3.14, the reader will infer that the true value lies between 3.135 and 3.145. If you're pretty sure it's between 3.138 and 3.142, it's best to write 3.140 ± 0.002.

How can you tell how many digits in your answer are significant? The foolproof method is the one I used for the circumference of the circle: calculate the result using all the possible values of the input numbers. There's a whole mathematical apparatus for doing this more efficiently, but we don't need to get into that here. Very crudely, if you used n significant digits in your input, you deserve n significant digits in your output.

> **Problem 6.** A notorious difficulty arises when you subtract two very nearly equal numbers. Here's an example. A machinist measures the lengths of two rods to fantastic precision: rod A is 4.793 02 m long, and rod B is 4.793 03 m long. (a) How many significant digits are there in each measurement? (b) What is the *difference* in their lengths? (c) How many significant digits are there in the difference?

1.2 Mechanics

Mechanics is **the study of motion**. It falls naturally into two parts: the *description* of motion (known technically as **kinematics**), and the *causes* of motion (**dynamics**).

1.2.1 Kinematics

Imagine a locomotive, constrained to move along a smooth, straight track.

We need some terminology to describe its motion.

- **Position.** First of all, how might I communicate to someone the *location* of the locomotive? One thing I could do is specify its distance from the lamppost, x. If I tell you that $x = 12$ m, you will know that the locomotive is 12 meters to the right of the lamppost. What would it mean if I reported $x = -3$ m? I suppose that would indicate that it is 3 meters to the *left* of the lamppost. Good: x is the **position** of the object.
- **Velocity.** When the locomotive moves, its position changes, and I might like to know how *fast* it changes. **Velocity** is the *rate of change* of position – the distance traveled, divided by the time it took to get there. For example, suppose we use a stopwatch to measure the time, t (in seconds), and we observe that the locomotive goes from $x = 10$ m, at $t = 3$ s, to $x = 25$ m, at $t = 8$ s. The distance traveled is 15 m, and the time it took was 5 s, so the velocity is

$$v = \frac{15\,\text{m}}{5\,\text{s}} = 3\,\text{m/s}.$$

Notice how we arrived at those numbers. To get the distance traveled, we subtracted the initial position from the final position: $x_{\text{final}} - x_{\text{initial}}$; to get the elapsed time, we subtracted the initial time from the final time: $t_{\text{final}} - t_{\text{initial}}$. Evidently

$$v = \frac{x_{\text{final}} - x_{\text{initial}}}{t_{\text{final}} - t_{\text{initial}}}.$$

But this looks awfully cumbersome; there is a nice shorthand, using the Greek letter Δ (delta), which means "the change in" whatever comes next:

$$\Delta z = z_{\text{final}} - z_{\text{initial}},$$

(whatever z may be). Then

$$v = \frac{\Delta x}{\Delta t}. \tag{1.1}$$

This doesn't tell us anything we didn't already know – it just says it in a tidier way.

What would you conclude if I told you the velocity of the locomotive is *negative* (say, -3 m/s)? Evidently in that case it is moving to the *left*. Physicists use the word **speed** to denote the *magnitude* of the velocity, regardless of its direction. In this example we would say the speed is 3 m/s (no minus sign – speed is always positive).

Motion at constant velocity. Suppose the locomotive started from the lamppost at $t = 0$, and traveled to the right at a constant velocity of 3 m/s. What is its position after 1 s? $x = 3$ m, obviously. After 2 s? $x = 6$ m. After

3 s? $x = 9$ m; etc. What's the general rule here? Evidently you multiply the velocity (v) by the elapsed time (t). We can express this in a simple formula:

$$x = vt \quad \text{(for motion at constant velocity).} \tag{1.2}$$

That's assuming it started at the lamppost ($x = 0$); if it started out at x_0, then after a time t it would be at

$$x = x_0 + vt. \tag{1.3}$$

- **Acceleration.** What if the locomotive speeds up or slows down, so its velocity is *not* constant? The rate of change of velocity is what we call **acceleration**. Acceleration is to velocity as velocity is to position:

$$a = \frac{\Delta v}{\Delta t}. \tag{1.4}$$

Say (for instance) the train was going 3 m/s at $t = 7$ s, and it's going 9 m/s at $t = 10$ s. Then $\Delta v = 9$ m/s $- 3$ m/s $= 6$ m/s, $\Delta t = 10$ s $- 7$ s $= 3$ s, so

$$a = \frac{6 \text{ m/s}}{3 \text{ s}} = 2 \text{ m/s}^2.$$

(Notice that the units of acceleration are m/s^2.)

Motion at constant acceleration. Suppose the locomotive starts from rest (velocity zero) at the lamppost (position zero), at $t = 0$ (noon, say), and undergoes constant acceleration, a. After a time t its velocity will be

$$v = at \quad \text{(for motion at constant acceleration)} \tag{1.5}$$

(just like the position, under motion at constant velocity).

How about its *position*? This is tricky. You might be inclined to say $x = vt = (at)t = at^2$, but this is incorrect, because *most* of the time the velocity was *less* than the final value at. (After all, it started out with velocity *zero*; Eq. (1.2) applies only to motion at *constant* velocity.) What *should* the formula be? I think it is plausible (and if this were a regular physics textbook I would prove it) that we want the *average* velocity: not the initial value (0) and not the final value (at), but halfway between ($v_{\text{average}} = at/2$):

$$x = v_{\text{ave}}t = \tfrac{1}{2}at^2 \quad \text{(for motion at constant acceleration).} \tag{1.6}$$

That's assuming it started from rest, at the lamppost; if it was already going at velocity v_0, then after a time t its velocity would be

$$v = v_0 + at, \tag{1.7}$$

and if it started from x_0 it would now be at

$$x = x_0 + v_0t + \tfrac{1}{2}at^2 \tag{1.8}$$

(x_0 is where it began, $v_0 t$ is the distance it would have gone if it had maintained its initial velocity, and $(1/2)at^2$ is the extra distance it went because of the acceleration).

A familiar example of motion at constant acceleration is **free fall**. When you drop an object, it accelerates (downward, of course) at $a = 9.81 \text{ m/s}^2$ (the so-called **acceleration of gravity**, for which we use the letter g). Remarkably, all objects fall with the *same* acceleration, if we neglect the effect of air resistance.

> **Example 2.** (a) Drop a rock off a high tower. After 3 s, how fast is it going, and how far has it fallen?
>
> *Solution:*
>
> $$v = at = (10 \text{ m/s}^2)(3 \text{ s}) = 30 \text{ m/s};$$
> $$x = \frac{1}{2}at^2 = \frac{1}{2}(10 \text{ m/s}^2)(9 \text{ s}^2) = 45 \text{ m}.$$
>
> (I used $g = 10 \text{ m/s}^2$, instead of 9.8 m/s^2, because I'm lazy.)
>
> (b) The Golden Gate Bridge is about 80 m above the ocean (actually, it's more like 67 m, but the numbers work out nicer this way). If you drop a quarter, how long will it take to hit the water?
>
> *Solution:*
>
> $$80 = \frac{1}{2}(10)t^2 \Rightarrow t^2 = \frac{2 \times 80}{10} = 16, \text{ so } t = 4 \text{ s}.$$
>
> More precisely (but this still ignores air resistance)
>
> $$67 = \frac{1}{2}(9.81)t^2 \Rightarrow t^2 = \frac{2 \times 67}{9.81} = 13.7, \text{ so } t = \sqrt{13.7} = 3.7 \text{ s}.$$
>
> (If you are a fastidious person you will want to attach the appropriate units to every quantity, as you go along. Sometimes this is distracting, however, and I'm not always conscientious about it. But the final answer definitely needs its units.)

Circular motion. Acceleration involves *change* in velocity. Ordinarily, this means speeding up or slowing down. But a change in *direction* also constitutes acceleration. The classic example is circular motion – a ball tied to a string whirled around your head, a planet orbiting around the Sun, or an electron circling the nucleus of an atom. If it's going in a circle, it is accelerating, toward the center, in the amount

$$a_c = \frac{v^2}{r}, \tag{1.9}$$

where v is the speed and r is the radius of the circle. We call it **centripetal** ("center-seeking") acceleration.[1]

> **Example 3.** You're on a merry-go-round, sitting on a horse 5 m from the center. It takes 10 s to complete a revolution. What is your speed? What is your centripetal acceleration?
>
> *Solution:* For one revolution, $vt = 2\pi r$, so $v = 2\pi r/t = 2\pi(5)/10 = 3.14$ m/s. Your acceleration, then, is
>
> $$a_c = \frac{3.14^2}{5} = 2.0 \, \text{m/s}^2.$$

Just for comparison, that's about 1/5 the acceleration of gravity.

Of course, you *feel* thrown outward, but your acceleration is actually inward. That's like taking off in an airplane: you're accelerating forward, so you *feel* pushed backward, into the seat.

Problem 7. A car driving north at constant velocity on SE 39th passes Tolman (600 meters south of Woodstock) at 12:07 p.m. and reaches Holgate (2400 meters north of Woodstock) at 12:12 p.m. Find: (a) Δx (in meters), (b) Δt (in seconds), and (c) the velocity of the car (in meters per second).

Problem 8. An ocean liner makes a 3600-mile voyage in 8 days, 8 hours. What was its (average) velocity?

Problem 9. An object moving initially with a velocity of 12 m/s is uniformly accelerated at a rate of 3 m/s^2. What is its velocity after 8 seconds of acceleration? If it started at position 8 m, what is its position after 8 s?

Problem 10. An automobile starts from rest and after 3 seconds is moving with a speed of 21 m/s. If the acceleration was constant, how far did the automobile move in the first 2 seconds? How far did it move during the third second?

---

[1] If you're interested in seeing a *derivation* of the centripetal acceleration formula, look in any introductory physics textbook.

Problem 11. Suppose you drop a rock down a well, and 3 seconds later you hear the splash.
(a) How deep is the well? (That is: how far down is the surface of the water? Neglect air resistance and the time it took the sound to reach you.)
(b) How fast was the brick going when it hit the water?
(c) Sound travels at 340 m/s. If we *do* take into account the time it took the sound to reach you, what is the corrected depth of the well?

Problem 12. You're on an island at the equator when for some reason the Earth's rotation starts to speed up, until eventually (when $a_c = g$) gravity can no longer hold things down, and you have to tie yourself to a palm tree to keep from flying off. What is the length of a day (the time it takes the Earth to complete a full rotation) when this happens?

1.2.2 Dynamics

It's harder to move a heavy object than a light one. That commonplace observation is the basis of dynamics. I need to explain, precisely and quantitatively, what it means.

Let's begin with the concept of **mass**. Mass is a measure of the "amount of stuff" in an object. It is directly related to the more familiar notion of **weight**. We measure it in **kilograms** (kg) – one kilogram corresponds to 2.2 lbs, so if you weigh 132 lbs, your mass is 60 kg. But technically "weight" is the force exerted by gravity – in outer space (where there is no gravity) everything is "weightless." By contrast, your mass is the same wherever you go.

Aristotle taught that objects in motion naturally come to rest, unless somebody is there to keep pushing on them. That certainly sounds right. If I shove a book across the table, it does indeed come to a stop. What Aristotle didn't realize is that an unseen force is acting on the book: the force of **friction**. Galileo was apparently the first to understand that if you could get rid of the friction, the book would keep on moving. But because friction is ubiquitous,[2] Aristotle's claim seems plausible, and it took 1500 years to fix his error. Serious physics began with **Newton's first law** of motion.

Objects keep moving in a straight line, with constant velocity, unless acted upon by some force.

(Of course, if they start out at rest – velocity zero – they remain at rest.)

---

[2] Historically, planets, freely falling objects, and pendulums were about as close as you could get to frictionless motion, and that's one reason why they played such an important role in the early development of physics.

If a force *does* act, the velocity will change – the object will *accelerate*. The amount of acceleration is given by **Newton's second law** of motion:

$$F = ma. \tag{1.10}$$

Here F is the applied force, m is the mass, and a is the resulting acceleration. Newton's second law is the foundation for all of mechanics. In a sense, it subsumes the first law as a special case: if the force is zero, then the acceleration is zero, so the velocity is constant. Notice that mass is a measure of **inertia** – the object's *resistance to acceleration*. The greater the mass, the larger the force that will be required to achieve a given acceleration.

But what exactly is "force"? Well, it's a "push" or "pull." Ordinarily, the agency responsible is pretty obvious: a rope pulling on a wagon, the chain lifting the carcass of a wrecked car, my hand pushing a book across the table. But sometimes the force is not so visible – the gravitational force on a rock in free fall; the frictional force on a sliding book; the magnetic force that sticks things to the refrigerator. . . . Indeed, much of physics involves discovering the mechanism by which one object exerts forces on another.

Force is measured in newtons (N):

$$1\,N = 1\,kg\,m/s^2. \tag{1.11}$$

How big is a newton? Well, you can easily exert a force of 10 N with your little finger.

> **Example 4.** A 60 kg skater is pulled across a frozen lake[3] with a constant force of 120 N. What is her acceleration?
>
> *Solution:*
>
> $$120 = 60\,a \quad \Rightarrow a = \frac{120}{60} = 2\,m/s^2.$$
>
> Knowing the acceleration, we can go back to kinematics (Eqs. (1.5) and (1.6)) to figure out (for example) how fast she is going after 3 s ($v = at = 2 \times 3 = 6$ m/s), or how far she goes in 4 s ($x = (1/2)at^2 = (1/2)(2)(4)^2 = 16$ m).

> **Example 5.** What is the force of gravity on a rock (mass m) in free fall?
>
> *Solution:* Its acceleration is g, so
>
> $$F = ma = mg.$$

---

[3] Physicists love icy surfaces; it's a secret code, meaning "frictionless."

What if that rock is now sitting on the ground? It is no longer accelerating, so the force on it must be zero. Does that mean the force of gravity has somehow vanished? No! Gravity is still acting, but there is now a second force involved: the Earth exerts an *upward* force on the rock, which exactly cancels the downward force of gravity. The net force on the rock is zero, and that's why it doesn't accelerate.

The force of gravity (mg) on an object is what we call its "weight." Notice that weight is a *force*, so it is measured in newtons (or, in the English system of units, pounds). A 60 kg person weighs $60 \times 9.8 = 588$ N, on Earth, but on the Moon (where $g = 1.67$ m/s^2) the same person would weigh 100 N.

Forces act *between* objects: I push you; the speedboat pulls a water-skier; a magnet attracts a nail; the Earth attracts the Moon. Newton noticed a peculiar fact about all forces.

If object A exerts a force on object B, then object B exerts an equal and opposite force on object A.

(By "opposite" I mean opposite in *direction*: if A pushes B to the right, then B pushes A to the left.) This is **Newton's third law** of motion.

Newton's third law is very easy to misconstrue. Imagine a horse pulling (forward) on a cart (with a force F_{HC}). According to the third law, the cart pulls (backward) on the horse (F_{CH}). So why does anything move? Don't the two forces simply cancel? *No!* The two forces in Newton's third law act on *different objects:* one on B, and the other on A. The horse pulls forward on the cart, so the cart moves. Meanwhile the cart pulls backward on the horse, but the horse has another force acting on it: the *ground* is pushing it forward, with a force (F_{GH}) that is presumably greater than the backward force of the cart, so it too moves forward. *What?!* Who told the ground to push forward on the horse? Ah, well, you see, the old horse knows all about Newton's third law: he pushes backward on the ground (F_{HG}), knowing that the ground will then be obliged to push forward on *him*!

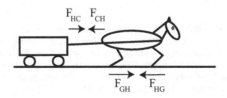

Or consider the case discussed in Example 5, where the rock is sitting on the ground. Gravity pulls it down, the Earth pushes it up; the two forces are equal and opposite, so the rock does not move. If you ask most physics students how they *know* that the two forces are equal, they will quote Newton's third law. It's not true; those two forces act on the *same object* (the rock), and Newton's third law has nothing to do with it. What the third law *does* say is that since the ground pushes up on the rock, the rock must push down on the ground (sure it does – that's why it leaves a dent in the dirt). And if the Earth pulls down on the rock (that's what the force of gravity is, as we shall soon see), then the rock pulls up on the Earth! How, then, *do* I know that the upward force of the ground exactly balances the downward force of gravity on the rock? Simply from the observed fact that the rock is not accelerating.[4]

> **Problem 13.** A 20 kg sled, starting from rest, is pulled across an ice-covered lake with a constant force of 5 N. (a) What is its acceleration? (b) How fast is it going after 4 seconds? (c) How far does it travel in the first 4 s?

> **Problem 14.** You are pushing a 60 kg trunk across the floor with a (horizontal) force of 450 N. The frictional force opposing the motion is 420 N. What is the acceleration of the trunk?

> **Problem 15.** A 1/2 kg book is shoved across the table with an initial velocity of 2 m/s. If it slides for 1 meter before coming to rest. (a) What was its acceleration? [Technically, this acceleration is negative, since the book is slowing down.] (b) What was the frictional force?

> **Problem 16.** Consider a 2 kg rock in free fall. The Earth is exerting a downward force mg on the rock, so according to Newton's third law the rock must be exerting an upward force mg on the Earth. Why don't we notice the Earth accelerating upward? [*Hint:* The mass of the Earth is 5.97×10^{24} kg. What is its acceleration?]

---

[4] OK, that's a cop-out. How did the ground know to push up with just exactly the right force? Well, think of the ground as a kind of "mattress." When you put the rock on it, the springs compress, and the more they compress the greater the upward force they exert on the rock. In fact, they squeeze down until their upward force exactly balances the downward force from the rock, and at that point there is no extra force to compress the springs any farther – they have automatically adjusted to match the weight of the rock!

1.3 Forces

Newton's second law ($F = ma$) tells us how an object (of mass m) will move (specifically, its acceleration, a), under the influence of a given force (F). But it doesn't tell us where that force came from in the first place – it doesn't say what agency or mechanism is responsible for it.[5] In this section I'll introduce the two most important forces of nature: the gravitational attraction between masses, and the electrical attraction or repulsion between charges.

Originally, I suppose, people thought forces could only exist between objects in **contact** – two colliding billiard balls, for example. That is perhaps why magnetic forces seemed so magical; two north poles repel each other even when they are not touching. Newton introduced the notion of **action-at-a-distance**: the Earth exerts a gravitational force on the Moon even though there is nothing at all connecting them. Newton himself thought this was absurd – he regarded his theory as provisional, pending the discovery of the unseen "chain" that transmits the force from one to the other.

In the early nineteenth century, Michael Faraday (working with magnets) introduced the concept of a **field**. Each magnet produces in its vicinity a **magnetic field** (sort of like the odor of a skunk), and it is this *field* that exerts a force on the other magnet (just as the smell of a skunk repels – or perhaps, attracts – other animals). This avoids the problem of action-at-a-distance, providing Newton's invisible "chain" linking the Moon to the Earth (only in that case it's a **gravitational field**). The field formulation turned out to be extraordinarily powerful, and by the twentieth century physicists had come to regard *all* forces as mediated by fields.

1.3.1 Newton's law of universal gravitation

According to Newton's **law of universal gravitation**, any two objects attract one another with a force that is proportional to the product of their masses, and inversely proportional to the square of the distance (r) between them:

$$F = G \frac{m_1 m_2}{r^2}. \tag{1.12}$$

---

[5] To my eye, $F = ma$ is logically confusing, because it *looks* like a formula for F, whereas in practice it is a formula for a. I think it would be preferable to write it in the (mathematically equivalent) form

$$a = \frac{F}{m}.$$

But everyone does it the other way, so I guess we had better stick with the bad notation.

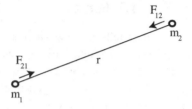

The proportionality factor G is a fundamental constant of nature:[6]

$$G = 6.674 \times 10^{-11}\,\mathrm{N\,m^2/kg^2}. \tag{1.13}$$

Notice that the force of m_1 on m_2 is the same as the force of m_2 on m_1 – and opposite in direction (since they are both attractive). So the law of universal gravitation is consistent with Newton's third law (he'd have been in serious trouble, of course, if it had *not* been so).

> **Example 6.** What's the force of (gravitational) attraction between you and me, if we're 2 m apart? Say you are 60 kg, and I'm 80 kg.
>
> *Solution:*
>
> $$F = (6.67 \times 10^{-11})\frac{60 \times 80}{2^2} = 8 \times 10^{-8}\,\mathrm{N}.$$
>
> Pretty feeble (nothing personal – gravity just *is* very weak).

What, then, is the net gravitational force on Newton's proverbial apple, at the surface of the Earth? When you stop to think about it, this is not an easy problem. After all, the apple is attracted by every rock and every iceberg, every cabbage in Nebraska and every kangaroo in Australia, and all the unknown stuff down there at the center of the Earth – these things are at all different distances, and they're not even pulling in the same direction.

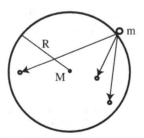

---

[6] That is, it cannot be derived from something else; it just is what it is.

It is an astonishing fact that when you add it all up, a spherical object (like the Earth) it acts *as if its mass were concentrated at the center.*[7] Thus the force on Newton's apple (mass m) is

$$F = G\frac{Mm}{R^2},$$

where M is the total mass of the Earth (5.97×10^{24} kg) and R is the radius of the Earth (6.37×10^6 m). In particular, the *acceleration* of Newton's apple (if it's in free fall) is, according to the second law,

$$a = \frac{F}{m} = \frac{1}{m}G\frac{Mm}{R^2} = G\frac{M}{R^2} = (6.67 \times 10^{-11})\frac{5.97 \times 10^{24}}{(6.37 \times 10^6)^2} = 9.81 \text{ m/s}^2,$$

which (of course) is precisely g, the acceleration of gravity. Notice that the mass of the *apple* canceled out; that's why all objects near the surface of the Earth fall with the *same* acceleration.[8]

Problem 17.

(a) Find the force of attraction between the Sun (mass 1.99×10^{30} kg) and the Earth (they are 1.50×10^{11} m apart).

(b) Find the force of attraction between the Moon (mass 7.36×10^{22} kg) and the Earth (they are 3.82×10^8 m apart).

Problem 18.

(a) Calculate the acceleration of gravity on the Moon (mass 7.36×10^{22} kg, radius 1.74×10^6 m).

(b) What would it be on Jupiter, whose mass is 318 times that of the Earth, and whose radius is 11 times as great?

1.3.2 Planetary motion

Imagine whirling a rock around over your head, on the end of a string. This is circular motion, so the rock has a centripetal acceleration. According to

---

[7] The proof of this is not easy, and Newton himself struggled (unsuccessfully) to find a simple argument. He knew it was true, by using calculus, but since that, too, was his own recent invention, not everyone found it persuasive.

[8] Note that "big-G" (Newton's constant) is a universal number – the same everywhere. But "little-g" (the acceleration of gravity) depends entirely on where you are. It even varies (slightly) from place to place on the surface of the Earth. It's different at the north pole than at the equator (the Earth isn't perfectly spherical), and lower at the top of Mount Everest (farther from the center). In fact, minute changes in g can be used to detect oil or mineral deposits underground.

Newton's second law, acceleration requires force. In this case the force on the rock is exerted by the string (and ultimately by your hand, pulling on the other end). If the string breaks, the force is gone, and the rock flies off in a straight line (according to the *first* law), tangential to its original circular trajectory.[9] It was the tension in the string, constantly pulling the rock in toward the center, that sustained the circular motion.

Planets travel in roughly circular orbits, but this time it is the gravitational attraction of the Sun that plays the role of the string.[10] The force on the planet (mass m) is $F = GMm/r^2$ (where M is the mass of the Sun and r is the radius of the orbit), and the centripetal acceleration is $a = v^2/r$, so Newton's second law ($F = ma$) says

$$G\frac{Mm}{r^2} = m\frac{v^2}{r}, \quad \text{or} \quad v^2 = \frac{GM}{r} \tag{1.14}$$

(the ms cancel, as does one factor of r). This is a remarkable result – it tells us the speed (v) of the planet, if we know the radius of its orbit (r). For instance, the distance to the Sun is $r = 1.496 \times 10^{11}$ m and the mass of the Sun is $M = 1.988 \times 10^{30}$ kg, so the speed of the Earth in its orbit is

$$v = \sqrt{\frac{GM}{r}} = \sqrt{\frac{(6.674 \times 10^{-11})(1.988 \times 10^{30})}{1.496 \times 10^{11}}} = 2.98 \times 10^4 \text{ m/s}.$$

Now, the speed of a planet is related to its **period** (T) – the time it takes to complete one revolution (a "year," for the planet in question). The *distance* it travels is the circumference of the orbit:

$$vT = 2\pi r, \tag{1.15}$$

so

$$\sqrt{\frac{GM}{r}}\, T = 2\pi r, \quad \text{or} \quad T = 2\pi r\sqrt{\frac{r}{GM}} = 2\pi\sqrt{\frac{r^3}{GM}}. \tag{1.16}$$

For the Earth,

$$T = 2\pi\sqrt{\frac{(1.496 \times 10^{11})^3}{(6.674 \times 10^{-11})(1.988 \times 10^{30})}} = 3.156 \times 10^7 \text{ s} = 365 \text{ days}.$$

[9] That's right – *tangential*, not radial – the same direction it was going when the string broke. If you doubt me, try it (but watch out for the windows).

[10] I don't know who was the first person to realize that uniform circular motion involves acceleration. Perhaps it was Newton himself. At any rate, it was a tremendously important conclusion, for acceleration requires *force*, and in the case of planetary motion it was Newton's search for the force responsible that led him to the law of universal gravitation.

Of course, we already knew that – but now we can calculate the length of a "year" for *any* planet. All we need to know is its distance from the Sun.[11]

Problem 19.

(a) The radius of Mercury's orbit is 5.79×10^{10} m. Calculate the length of a "year" on Mercury.

(b) Mars is 2.28×10^{11} m from the Sun. How long does it take Mars to complete one revolution?

Problem 20. Calculate the period of the Moon's orbit around the Earth. (The mass of the Earth is 5.97×10^{24} kg, and the distance to the Moon is 3.82×10^8 m.) Give your result in days.

Problem 21. A "synchronous" satellite is one that remains above the same point on the equator at all times. (a) What is the period of a synchronous satellite? (b) How high above the surface must it be?

1.3.3 Coulomb's law

Electric charge is the stuff that makes your hair stand on end when you comb it on a dry winter day, or produces a spark when you drop a screwdriver across the terminals of a car battery (and more dramatically the bolts of lightning in a thunderstorm). Electrical forces account for most of chemistry (and, presumably, for all of biology). Moving electric charges (**currents**) make lightbulbs bright and irons hot; they are responsible for all of electronics. Magnetism, too, is due to electric charges in motion. Indeed, it is no exaggeration to say that we live in an electrical world, for with the exception of gravity every force you encounter in daily life is electrical in nature (unless you work at a nuclear power plant).

Unlike mass, which is always positive, electric charge can be positive or negative, and opposite charges cancel, in the sense that if you put a negative charge on top of an equal positive charge, the combination is electrically neutral. Electric charge is measured in coulombs (C). The charge of an electron is

---

[11] By meticulous astronomical observation, Kepler arrived at three laws of planetary motion, one of which says that the square of the period (T^2) is proportional to the cube of the radius (r^3). Newton's law of universal gravitation *explained* Kepler's law, and provided a formula for the constant of proportionality ($4\pi^2/GM$).

-1.602×10^{-19} C, and the charge of a proton is $+1.602 \times 10^{-19}$ C; atoms (which contain equal numbers of electrons and protons) are electrically neutral. A typical household appliance takes an electric current of (say) one ampere (1 C/s), which means that roughly 10^{19} electrons per second pass any given point in the wire.

Like charges repel, and opposite charges attract, with a force that is proportional to the product of the charges (q_1 and q_2), and inversely proportional to the square of the distance between them. This is called **Coulomb's law**:

$$F = k \frac{q_1 q_2}{r^2}. \tag{1.17}$$

The proportionality factor k is another fundamental constant of nature,

$$k = 8.988 \times 10^9 \, \mathrm{N\,m^2/C^2}. \tag{1.18}$$

Coulomb's law is strikingly similar to Newton's law of universal gravitation. But on a fundamental level, electrical forces are astronomically more powerful than gravitational forces. For instance, two electrons (mass $m = 9.109 \times 10^{-31}$ kg) attract one another gravitationally, but repel one another electrically. The ratio of the two forces is

$$\frac{F_{\text{elec}}}{F_{\text{grav}}} = \frac{kq^2/r^2}{Gm^2/r^2} = \frac{kq^2}{Gm^2} = \frac{8.988 \times 10^9 (-1.602 \times 10^{-19})^2}{6.674 \times 10^{-11}(9.109 \times 10^{-31})^2} = 4.17 \times 10^{42}.$$

(Notice that the rs canceled out, so it doesn't matter how far apart they are – the *ratio* of the forces is always the same.) This is a number so enormous – 4 followed by 42 zeros – it is hard to comprehend.

That being the case, how come we aren't more conscious of electrical forces in everyday life? We are certainly well aware of gravity. The answer is that bulk matter is electrically neutral, to fantastic precision – it contains equal amounts of positive and negative charge. In your body there are about 10^{28} protons, and an equal number of electrons. Walk across a fuzzy rug, and you might pick up an extra 10^{11} electrons (which drain off, giving you a nasty shock, when you touch a metal doorknob). That may sound like a lot, but as a fraction of the total it is only $10^{11}/10^{28} = 10^{-17}$, or 0.000 000 000 000 001%. Even when charged up, you are, in absolute terms, very close to electrically neutral.

Problem 22. For a lecture demonstration two wads of paper are charged up, one to $+3 \times 10^{-8}$ C, and the other to -4×10^{-8} C. They are suspended by thin threads and held 2 cm (0.02 m) apart. What is the force of attraction between them?

Problem 23. Suppose you could remove just 0.1% of the electrons from two people, so their net charge is that of 10^{25} protons. What would the force of repulsion between them be, if they stood 2 m apart?

1.4 Conservation laws

You and some friends get together for an evening of poker. At the beginning everyone buys $25 worth of chips. In the course of the evening George loses it all, the rest of you win or (more often) lose modest amounts, and Sarah cleans up, going home with $103. The details depend on skill and luck, of course, but this much I can say for sure: unless someone is seriously cheating, the *total* amount of money at the end is the same as it was at the beginning – it has been redistributed among the players, but none has been created or destroyed.

This is a model for what physicists would call a **conservation law**: some quantity is the same (in total) at the end of a process as it was at the beginning. The most familiar conservation law in classical mechanics is the **conservation of mass**: you can slice a potato into 17 pieces, or melt an ice cube, or pack together some lumps of clay, but in every case the total mass is unchanged.[12] Two other such laws are **conservation of momentum** and **conservation of energy**. I'll discuss those in the next few sections.

Conservation laws can be extraordinarily powerful and efficient. If all you're interested in is the total amount of cash (in my example), and you don't care whose wallet it is in, then you don't need to know anything about the game itself (the *mechanism* by which the money gets from one player to another).

Example 7. A 2 cm × 2 cm × 2 cm cube of ice melts. What is the volume of the puddle?

Solution: The ice cube is $2^3 = 8$ cubic centimeters, and the density of ice is 0.917 grams per cubic centimeter, so its mass is $8 \times 0.917 = 7.336$ grams. Conservation of mass says this must also be the mass of the water, and the density of water is exactly[13] 1 gram per cubic centimeter, so there will be 7.336 cubic centimeters of water in the puddle. Notice that the volume of water is *not* the same

---

[12] Of course, you've got to count *all* of the mass. If you ate one of the pieces of potato, or you allowed some of the water to evaporate, or you left some of the clay on the table, then it may appear that mass was not conserved, but that was just carelessness on your part – the law of physics is correct.

[13] This is not a coincidence – it's how the gram was defined.

as the volume of ice – if you poured the water back into the form
that made the ice cube, it would not completely fill it. There is no
law of conservation of volume!

1.4.1 Momentum

Momentum is a measure of "how hard it is to stop something" – how much
padding you need on the catcher's mitt. Obviously, it depends on the velocity
of the object (a 90 mph pitch will be harder to stop than one going 30 mph),
and pretty clearly it will also depend on its mass (it's easier to stop a ping-pong
ball than a golf ball). The official definition of momentum (for which we use
the letter p) is "mass times velocity":

$$p = mv. \tag{1.19}$$

There are no special units for momentum – it's just kg m/s.

The law of **conservation of momentum** says that in any process the sum
of the momenta (of all the objects involved) is the same afterward as it was
before:

$$p_{\text{final}} = p_{\text{initial}}. \tag{1.20}$$

There are many applications of conservation of momentum; the most common
is to **collisions**.

> **Example 8.** A 3 kg lump of clay going 4 m/s collides with a 1 kg
> lump at rest, and they stick together. What is the velocity of the
> resulting composite lump, immediately[14] after the collision?
>
> *Solution:* The total initial momentum was $[(3)(4) + (1)(0)] = 12$,
> and the final mass is $(3 + 1) = 4$ kg, so the final momentum is
> $(4)(v)$, where v is the (unknown) final velocity. Conservation of
> momentum says $4v = 12$, so $v = 3$ m/s.

Notice that I never had to inquire about the actual forces involved – the
great virtue of conservation laws is that they step right over the messy and
complicated *details* of the process, and go straight to the relatively simple final
result.

---

[14] Of course, if you wait half an hour all bets are off – other things (friction, gravity, whatever)
come into play, and they gobble up some of the momentum. It's best to think of a "collision"
as something that happen so fast that only the impulse between the objects has time to affect
their motion. Or, if you prefer, imagine that it occurs in outer space, where there are no other
influences.

Problem 24. A baseball (mass 0.142 kg) and a car (mass 1000 kg) are both traveling at 45 m/s (that's about 100 mph). Find the momentum of each object.

Problem 25. On a frictionless racetrack a model car of mass 5 kg going at 9 m/s rear-ends a car of mass 3 kg going in the same direction at 1 m/s. If they stick together, what are the mass and velocity of the combination? What if they were going in *opposite* directions before the collision?

Problem 26. A 4000 kg truck going 30 m/s rear-ends a 1000 kg VW bug going 10 m/s. After the collision, the truck is going 29 m/s. How fast is the VW going?

1.4.2 Work and power

Physicists use a lot of words that are familiar in everyday life – speed, velocity, acceleration, force, momentum, ... But in physics they become technical terms, with precise definitions. Usually the technical meaning amounts to a refinement of the casual usage, and one's reaction is (for example), "Yes, I guess that's what I always meant by 'acceleration'." But occasionally the correspondence is not so obvious, and you need to be more careful: intuitions (based on the everyday use of the word) may not hold in the technical sense. "Work" and "energy" are two particularly slippery cases, and a lot of nonsense in the press and popular culture results from people confusing the technical term with its everyday counterpart.

In physics, **work** is defined as *force* times *distance*:

$$W = Fd. \tag{1.21}$$

I think this makes sense. If you push a heavy trunk across the floor, you have surely done some work; push twice as hard (or move it twice as far) and it seems reasonable that you have done twice as much work. The units of work are **joules** (J): $1\,\text{J} = 1\,\text{N}\,\text{m} = 1\,\text{kg}\,\text{m}^2/\text{s}^2$.

Example 9. How much work does it take to lift a brick (mass 2 kg, say) from the floor to a shelf (height 1.5 m, say)?

Solution: To lift it you must exert an upward force mg, counteracting the force of gravity (of course, you could exert *more* than this, if you're impatient, but we're interested in the *minimum* amount of work it takes). So the work you must do is

$$W = Fd = (mg)(h) = (2)(10)(1.5) = 30\,\mathrm{J}.$$

So far, so good. But notice that you have to *move* an object, to do any work on it (if $d = 0$ then $W = 0$). If you simply *hold* the brick in mid-air for half an hour, you may be sweating and straining, but you're not doing any work. You might as well have put it on the shelf; the shelf isn't doing any work, and neither are you, if you insist on taking its place. What is more, if you *lower* the brick from the shelf back to the floor, you are (technically) doing *negative* work ($-30\,\mathrm{J}$, in this case), because the force you exert (up) and the displacement (down) are now in opposite directions. You wouldn't really have to do work at all, to get the brick to the floor – you could just *drop* it, and let it do some work for *you* (pound in a nail, for example).

Finally, motion perpendicular to the force doesn't count. If you carry a brick around the room, you're not doing any work, because the force you exert (up) is perpendicular to the motion (horizontal). If you were smart, you'd put it on a (frictionless)[15] cart, and give the cart a slight tap to get it going; the cart doesn't do any work as it transports the brick around, and neither do you, if you decide to carry it yourself. It's only when you lift the brick up or lower it down that you do work.

Notice the syntax: work is something you *do* on an object. It doesn't have to be *you*, of course – any agency that exerts a force can do work. In the Example 9, for instance, gravity did (negative) work on the brick, while you did an equal amount of positive work. When the brick is lowered back down, gravity does *positive* work, since the force of gravity (down) is now in the same direction as the displacement. So you have to careful: to determine how much work is being done, you need first to ask "by *whom?*"

Power, is the *rate* at which work is performed:

$$P = \frac{\Delta W}{\Delta t}. \tag{1.22}$$

It is measured in J/s, or **watts** (W). For example, if it took you 2 seconds to lift the brick (in Example 9), your power output was $30/2 = 15\,\mathrm{W}$, the same as a dim light bulb.

Problem 27. You need to move a trunk across the floor, a distance of 5 meters. If you push with a force of $200\,\mathrm{N}$, how much work must you do? If it takes you 10 seconds, what was your power output?

---

[15] If there is friction, then you'll have to push on the cart, of course, to keep it going, and your *horizontal* force does do work.

Problem 28. Suppose you carry a 50 kg sack of potatoes up two flights of stairs, a total height gain of 10 m. How much work did you do? If it took you 20 s, what was your power output?

Problem 29. A company makes electric generators out of converted bicycles. They claim that a good cyclist can generate 250 watts per hour. Explain why this is nonsense. What should they have said?

1.4.3 Energy

Energy is the *capacity to do work*. More precisely, the energy of a system is the amount of work it is capable of doing. (Since it's an amount of work, it is measured in the same units: joules.) This is sort of consistent with our everyday use of the word: if I wake up "full of energy," that means I am ready to do some work. But when practitioners of Chinese medicine speak of "fields of energy," or physicists (who ought to know better) tell you that "mass is energy," they are misusing the language. Mass doesn't even have the same *units* as energy, and I don't know *what* the good doctors are talking about, but it certainly isn't energy, in the technical sense.

Energy comes in many different forms. For example:

- **Thermal energy:** the capacity to do work by virtue of temperature – hot water can do more work for you than cold water.
- **Chemical energy:** the capacity to do work by virtue of chemical composition – a stick of dynamite can do more work than a banana. (Coal, oil, and natural gas are all repositories of chemical energy; so are batteries.)
- **Electrical energy:** separated electric charges can do work by virtue of the Coulomb force between them.

But for our purposes the two most important forms of energy are kinetic and potential.

- **Kinetic energy:** the capacity to do work by virtue of *motion*. A wrecking ball swinging from a crane can do some work for you that the same ball just hanging at rest could not. Like momentum, kinetic energy obviously depends on mass and velocity; it turns out that the precise amount of work a moving object is capable of doing is

$$T = \tfrac{1}{2}mv^2. \tag{1.23}$$

- **Potential energy:**[16] the capacity to do work by virtue of *location*. A brick four feet above the floor is capable of doing some work for me (I could drop it, and have it pound in a nail), which the same brick resting on the floor is not able to do. Specifically, the potential energy of an object of mass m at a height h is

$$V = mgh. \tag{1.24}$$

Wait a minute! Height above *what*? Good question! *Answer:* It doesn't matter. Only *changes* in potential energy carry physical significance. If you're planning to drop the object to the floor, then it would be convenient to measure h up from the floor, but if you are thinking of dropping it out the window, then the distance to the ground would be more relevant – it's the distance it falls ($h = h_{\text{initial}} - h_{\text{final}}$) that tells you how much work it will do. Potential energy is like altitude, in this respect. If I tell you that Denver is at 5280 feet, you will naturally assume I mean above sea level, because that is the customary reference point, but we could just as well have measured altitudes above Washington, DC, for example, or the bottom of Death Valley. On the other hand, if I tell you that the *difference* in altitude between Denver and the summit of Pike's Peak is 8830 feet, that's a physical fact independent of anybody's convention. Usually, a sensible reference point for potential energy suggests itself ("the floor," "ground level," or whatever).

Example 10. What is the kinetic energy of a baseball (0.142 kg) going 90 mph (40 m/s)?

Solution:

$$T = \tfrac{1}{2}(0.142)(40)^2 = 114 \,\text{J}.$$

Example 11. How much work could a 0.6 kg book do, if you dropped it from your desk to the floor (1 m below)? What if you dropped it out the window, to the ground 20 m below?

Solution:

$$V_{\text{floor}} = (0.6)(9.8)(1) = 6\,\text{J}; \quad V_{\text{ground}} = (0.6)(9.8)(20) = 120\,\text{J}.$$

Problem 30. (a) What is the kinetic energy of an athlete ($m = 75$ kg), running a 100 meter dash in 10 seconds? (Assume constant speed.) (b) What is his momentum?

---

[16] I hate that term, because it makes it sound as though "potential" energy is in some sense not (yet?) "real." I suppose there was a time when kinetic energy seemed "genuine," and potential energy, being less tangible, was given this misleading name for contrast.

Problem 31. You are pulling a child on a sled across a frozen pond (the ice makes it frictionless). The mass of the child-plus-sled is 20 kg, and you exert a force of 40 N for 3 seconds, starting from rest.

(a) What is the acceleration of the sled?
(b) How far does the sled go?
(c) How fast is it going, at the end?
(d) How much work did you do?
(e) What is the final kinetic energy of the sled-plus-child?
(f) What was your power output?

Problem 32. At full capacity 11 000 000 kg of water flow over Hoover dam every second. It falls 220 m to the turbines below. If all of its potential energy is converted into work, what is the power generated?

1.4.4 Conservation of energy

Energy cannot be created or destroyed. In any process, the *total* energy (of all forms, and for all the objects involved) is the same afterward as it was before:

$$E_{\text{final}} = E_{\text{initial}}. \tag{1.25}$$

This is the **law of conservation of energy**. You can convert energy from one *form* to another (in free fall potential energy is converted into kinetic energy), and you can transfer it from one *object* to another (in the collision of two billiard balls, kinetic energy is passed from one to the other), but the sum total never changes.

In mechanics the relevant forms of energy are kinetic and potential, so conservation of energy takes the form

$$T_{\text{final}} + V_{\text{final}} = T_{\text{initial}} + V_{\text{initial}}. \tag{1.26}$$

However, friction typically converts kinetic energy into thermal energy (heat). For instance, when you shove a book across the table, and friction brings it to a stop, the temperature of the book rises slightly. Because the energy is gone, as far as mechanics is concerned, physicists call friction a "nonconservative" force. I think that's misleading; the energy is still there, it's just been converted into a form that is no longer relevant to the motion of the object.

Conservation of energy offers miraculously quick and easy solutions to many problems you could have solved more laboriously by applying Newton's second law.

Example 12. Suppose you drop a rock off a cliff 20 m high. How fast is it going when it reaches the bottom? [Pretend that $g = 10 \text{ m/s}^2$, so the numbers work out nicely.]

Solution: It started with potential energy and no kinetic energy, and it ended up with kinetic energy and no potential energy. Conservation of energy requires that

$$\tfrac{1}{2}mv^2 = mgh \text{ or } v^2 = 2gh = 2(10)(20) = 400 \Rightarrow v = 20 \text{ m/s}.$$

Of course, you could have done this the "old fashioned" way: the acceleration of the rock is constant ($a = 10 \text{ m/s}^2$), so the distance it travels in time t is (Eq. (1.6)) $(\tfrac{1}{2})at^2$. It falls a distance 20 m, so

$$\tfrac{1}{2}(10)t^2 = 20 \quad \Rightarrow \quad t^2 = 4 \quad \Rightarrow \quad t = 2 \text{ s}.$$

That's the *time* it takes, so its velocity (Eq. (1.5)) will be

$$v = at = (10)(2) = 20 \text{ m/s}.$$

Same answer, but notice how much quicker conservation of energy was – I never had to calculate the *time* it took to fall. (On the other hand, if t is what you're really interested in, conservation of energy won't help you.)

By definition, energy is the amount of work the system can do for you. Because energy is conserved, it is also the amount of work you had to do to assemble the system in the first place. In the meantime it is "in the bank," waiting to be put to use. You do some work on a system (lift the brick up to a shelf, say) – that's like a "deposit"; later on you let the system do some work for you, returning to its initial state (drop the brick) – that's like a "withdrawal." But at this bank there are no fees and you draw no interest (that's conservation of energy) – you get out precisely what you put in. Energy, then, is "stored work."

Example 13. By applying a constant force to an object of mass m, you accelerate it up to speed v. How much work did you do?

Solution: Since the force is a constant, so too is the acceleration ($F = ma$), and therefore the distance traveled is $d = \tfrac{1}{2}at^2$. Therefore the work done is

$$W = Fd = (ma)\left(\tfrac{1}{2}at^2\right) = \tfrac{1}{2}m(at)^2 = \tfrac{1}{2}mv^2,$$

which is precisely the kinetic energy now stored in the moving object (the amount of work it is now capable of doing for you).

Example 14. How much work does it take to lift an object of mass m to a height h?

Solution: We already calculated this in Example 9:

$$W = Fd = (mg)(h) = mgh,$$

which is precisely the potential energy now stored in the object (the amount of work it is now capable of doing for you).

Problem 33. A rock is thrown straight up with an initial velocity of 30 m/s. How high will it rise before coming back to Earth?

Problem 34. An orange is dropped from the observation deck of the Empire State Building (320 meters high). How fast is it going when it hits the sidewalk (neglecting air resistance)?

Problem 35. A lump of clay is thrown at a brick wall, and sticks. (a) What happened to its momentum? (b) What happened to its kinetic energy?

Problem 36. The thermal energy of an object is cmT, where c is the "specific heat" of the substance, m is its mass, and T is its Celsius[17] temperature. Suppose you slide a book across the table, with an initial speed of 5 m/s. Friction brings it to rest, converting its kinetic energy into thermal energy. If its specific heat is 1000 J/(kg °C), how much does its temperature rise?

1.4.5 Gravitational potential energy

The formula $V = mgh$ is all very well for objects close to the surface of the Earth, where g is a constant 9.81 m/s^2, but what if you go *way* far out, where the force of gravity is weaker? What is the gravitational potential energy of an object of mass m at a distance r from the center of a body (such as the Earth) of mass M? It simplifies matters to change the reference point, and ask, "How much work would it take to bring m in from infinitely far away?" This

---

[17] Actually, it's the "Kelvin" temperature (the temperature above absolute zero), but as long as you're interested only in *changes* in temperature, it doesn't matter which one you use.

28 *Introduction: classical foundations*

calculation requires integral calculus,[18] but the result is very simple:

$$V = -G\frac{Mm}{r}.\tag{1.27}$$

Why *negative*? Well, m is *attracted* by M, so you don't have to push on it at all – on the contrary, you could harness it up and have it do work for *you*. To put it the other way around, if you want to *disassemble* the system, taking m back out to "infinity," you will have to do work on it, fighting against the gravitational attraction of M.[19] The same idea would apply to a rock at the bottom of a deep hole; then h would be negative, and the potential energy (mgh) would be negative, and this reflects the fact that it can't do work for you – on the contrary, you would have first to lift it up out of the hole, just to get it to ground level.

> **Example 15.** Suppose you dropped a rock from a height above the surface equal to the radius of the Earth (so the distance from the center is $r = 2R$). How fast would it be going when it hit the surface?
>
> *Solution:* In the beginning the rock had potential energy
>
> $$-G\frac{Mm}{2R},$$
>
> but no kinetic energy; at the end it had both potential energy ($-GMm/R$) and kinetic energy ($\frac{1}{2}mv^2$). Conservation of energy says
>
> $$-G\frac{Mm}{2R} = -G\frac{Mm}{R} + \frac{1}{2}mv^2 \Rightarrow v^2 = -G\frac{M}{R} + 2G\frac{M}{R} = G\frac{M}{R},$$
>
> so
>
> $$v = \sqrt{\frac{GM}{R}} = \sqrt{\frac{(6.67 \times 10^{-11})(5.97 \times 10^{24})}{6.37 \times 10^6}} = 7900\,\text{m/s}.$$
>
> Conversely, if you throw a rock straight up from the surface of the Earth at 7.9 km/s, it will rise to a height equal to the radius of the Earth before turning around and falling back.

This begs an interesting question: could you throw a rock fast enough that it would *never* come back, but would continue all the way out to infinity? Well,

[18] Because the force changes, as m gets closer to M, you must chop the path into short segments and apply the definition ($\Delta W = F\,\Delta d$) to each one.

[19] If we had used the surface of the Earth as our reference point, then the potential energy would have come out positive – and the formula would be more complicated. But remember: only *differences* in potential energy matter, and they are the same regardless of the reference point.

in that case you would start with potential energy $(-GMm/R)$ and kinetic energy $(\frac{1}{2}mv^2)$; at the end of the trip the potential energy would be *zero* (since r is infinite), and the kinetic energy would also be zero (no point in wasting energy – we want the critical case where the rock just barely makes it out there; throw it even harder, and it will "arrive" with kinetic energy to spare). So conservation of energy says

$$-G\frac{Mm}{R} + \frac{1}{2}mv^2 = 0, \quad \Rightarrow \quad v = \sqrt{\frac{2GM}{R}}. \qquad (1.28)$$

For the Earth, the **escape velocity** is

$$v = \sqrt{\frac{(2)(6.67 \times 10^{-11})(5.98 \times 10^{24})}{6.37 \times 10^6}} = 1.12 \times 10^4 \,\text{m/s}. \qquad (1.29)$$

Because Coulomb's law has the same mathematical structure as Newton's law of universal gravitation, we can immediately write down the *electrical* potential energy of a charge q at a distance r from a stationary charge Q:

$$V = k\frac{qQ}{r}. \qquad (1.30)$$

This time there is *no* minus sign, because like charges *repel*, and you would have to do work to bring them together (of course, if they are *opposite* charges you pick up a minus sign from the product qQ).

Problem 37. What is the escape velocity of the Moon? (Refer to Problem 18 for the mass and radius.)

Problem 38.

(a) Suppose you took the Earth, and (without changing its mass) squeezed it down until its escape velocity was equal to the speed of light (3.00×10^8 m/s). Then even light could not escape, and the Earth would be a **black hole**. What would the Earth's radius be, in that case (in meters)?

(b) What would the radius of the Sun be, if it were squeezed down to a black hole?

Problem 39. Show that the kinetic energy of a planet is half as great as its potential energy (because the potential energy is *negative*, so too is the *total* energy).

1.5 Waves

Suppose you had a long rope;[20] you're holding one end, and the other is tied to a tree (say). Shake your end up and down rhythmically, and a wave travels down the line.[21]

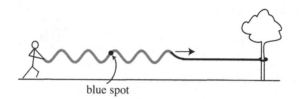

blue spot

Part way along somebody has put a dab of blue paint on the rope. *Question:* How does that blue spot move, as the wave passes by? Naively, you might expect that it would be swept along by the wave. But that can't be right – after all, the spot is at a fixed location on the rope. In fact, the *spot* simply moves up and down, while the *wave* moves to the right. The wave is a kind of epiphenomenon – a *pattern* arising from the collective motion of different points on the rope. It is the *shape* of the rope, not the rope itself, that travels to the right.

1.5.1 Velocity, wavelength, and frequency

We need some terminology for describing waves.

- The **amplitude** of the wave, A, is a measure of how "big" it is – the height of a crest.
- The **speed** (or velocity) of the wave, v, tells you how fast it is traveling. Please note that this is the speed of a point *on the wave* – a crest, for instance – and is not to be confused with the speed of a point *on the rope* – the blue spot, for example – as it oscillates up and down.

- The **wavelength**, λ (Greek letter "lambda"), is the distance between adjacent crests.

---

[20] Actually, a slinky works better.

[21] When the wave gets to the tree it will "reflect" back, and now you've got *two* waves, propagating in opposite directions. I'll discuss that case in a moment, but for now let's keep it simple: the rope is long, and the wave hasn't reached the tree yet.

Table 1.1. *The visible range.*

Frequency (Hz)	Color	Wavelength (m)
1.0×10^{15}	near ultraviolet	3.0×10^{-7}
7.5×10^{14}	shortest visible blue	4.0×10^{-7}
6.5×10^{14}	blue	4.6×10^{-7}
5.6×10^{14}	green	5.4×10^{-7}
5.1×10^{14}	yellow	5.9×10^{-7}
4.9×10^{14}	orange	6.1×10^{-7}
3.9×10^{14}	longest visible red	7.6×10^{-7}
3.0×10^{14}	near infrared	1.0×10^{-6}

- The **period**, T, is the time it takes a point on the rope – the blue spot, say – to execute one full oscillation. Because λ is the distance the wave travels in one cycle, and T is the time it takes,

$$v = \frac{\lambda}{T}. \tag{1.31}$$

- It is customary to work with the **frequency**, f, instead of the period (T). Frequency is the number of cycles per second, whereas period is the number of seconds per cycle; they are reciprocals:

$$f = \frac{1}{T}. \tag{1.32}$$

So I could rewrite the fundamental relation (Eq. (1.31)) in the more useful form

$$\lambda f = v. \tag{1.33}$$

The units of frequency are "cycles per second," or **hertz** (Hz).

All this applies to any kind of wave – waves on water,[22] sound waves,[23] radio waves, light, whatever. In the case of sound waves, frequency corresponds to *pitch*. The higher the frequency, the higher the pitch. For example, "concert A" is 440 Hz. The audible range for humans runs from about 16 Hz (a deep bass note) up to 20 000 Hz (a shrill squeak).

For light waves, frequency corresponds to *color*. Red light has a frequency around 4×10^{14} Hz, and blue light is around 6.5×10^{14} Hz. The visible range for humans runs from 3.9 to 7.5 $\times 10^{14}$ Hz (Table 1.1). This represents only

---

[22] In the case of water waves the actual motion of a droplet is not up-and-down, but circular – that's how a cork floating on the surface, for instance, would move. But never mind; the fact remains that every point in the medium executes a little dance in place, while the wave itself passes by.

[23] For sound waves the air molecules move forward and back along the direction of the wave, not up and down perpendicular to it, but once again there is no *net* displacement as the wave passes by.

Table 1.2. *The electromagnetic spectrum.*

Frequency (Hz)	Type	Wavelength (m)
10^{22}		10^{-13}
10^{21}	gamma rays	10^{-12}
10^{20}		10^{-11}
10^{19}		10^{-10}
10^{18}	X-rays	10^{-9}
10^{17}		10^{-8}
10^{16}	ultraviolet	10^{-7}
10^{15}	visible	10^{-6}
10^{14}	infrared	10^{-5}
10^{13}		10^{-4}
10^{12}		10^{-3}
10^{11}		10^{-2}
10^{10}	microwave	10^{-1}
10^{9}		1
10^{8}	TV, FM	10
10^{7}		10^{2}
10^{6}	AM	10^{3}
10^{5}		10^{4}
10^{4}	RF	10^{5}
10^{3}		10^{6}

a tiny "window" in a vast electromagnetic spectrum that goes from 10^4 Hz (radio waves) through microwaves, infrared, visible, ultraviolet, and X-rays, to gamma rays at 10^{22} Hz (Table 1.2). I shall use the word "light" as a generic term for all of them. Electromagnetic waves travel (in vacuum) at the speed of light, for which we reserve the letter c:

$$c = 2.998 \times 10^8 \, \text{m/s}. \qquad (1.34)$$

Problem 40. Suppose you shake the rope up and down twice a second. What is the period of the resulting wave? What is its frequency?

Problem 41. The speed of sound is 340 m/s. What is the wavelength of concert "A"?

Problem 42. Helium–neon lasers have a wavelength of 6.328×10^{-7} m. What is the frequency of this light? What color is it?

Problem 43. AM radio station KPOJ broadcasts at a frequency of 620 kHz
(6.2×10^5 Hz). What is the wavelength of the signal? What is the period of
the oscillations?

1.5.2 Interference

What happens when a wave comes to the end of the line (where the rope is tied
to the tree, for example)? It "reflects" back. Now we have *two* waves on the
rope – the original one propagating to the right, and the reflected one going to
the left.[24] What is the resulting shape of the rope? It is certainly not obvious,
but I will tell you the answer: the two waves pass right through each other –
the net displacement of the rope at any point (say, the blue spot) is the *sum*
of the displacements it would have had from each wave separately. This is
called the **principle of superposition**; it is the simplest behavior you could
possibly hope for, but its implications are astounding.

To understand exactly what it means, let's track two short pulses, approaching each other from opposite directions.

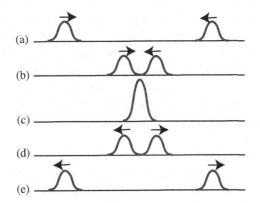

In scene (a) they are coming toward each other; in scene (b) they have just met;
and by scene (e) they have passed completely through and are moving apart.
In scene (c), where the two pulses momentarily coincide, the net displacement
is twice what it would have been for either one alone. Nothing too surprising
about that.

But what if one pulse is a valley, instead of a hill (shake the string
downward)?

---

[24] The same thing happens when water waves hit a barrier, sound echoes off a wall, or light
   strikes a mirror.

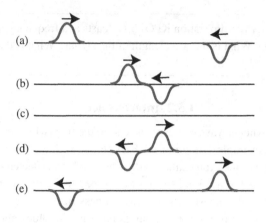

This time there comes an instant (c) when the two exactly cancel, and the string is momentarily straight – the two pulses add up to no pulse at all! We call this **destructive interference** (the previous case was **constructive interference**).[25]

Destructive interference is the unmistakable signature of a wave phenomenon. If you are talking about a rope, or water, where you can actually *see* the thing that's waving, there's no problem. But in the case of sound or light it's not so obvious that we're dealing with waves at all. Indeed, Newton thought light was a stream of *particles* ("corpuscles," he called them). How would you determine whether something is a wave or a particle, if you can't see it? *Answer:* You check for destructive interference. Two particles cannot add up to no particle at all, but two waves can combine to make no wave at all. In 1801, Thomas Young used destructive interference to prove that light is a wave phenomenon.

The basic idea is easy to demonstrate with water waves, using a "ripple tank." In a shallow tray of water, two small spheres bob up and down in unison, driven by a motor (in the figure they are 3λ apart, at the bottom edge). The point indicated toward the upper left is half a wavelength farther from the right sphere than the left one, so the waves arrive **out of phase** (one's a crest when the other is a trough), and they interfere destructively – at this point the water is flat. But a little to the left or right the waves are **in phase** (both crests, or both troughs) so they interfere constructively, and here the water is choppy. Notice that the points of destructive interference form lines ("nodal" lines). The nodal lines represent the locus of points that are $\lambda/2$, $3\lambda/2$, $5\lambda/2$, . . . farther from one sphere than the other.

---

[25] Personally, I think "interference" is precisely the *wrong* word for it. The two waves do *not* interact, and they emerge unscathed from their encounter. But I'm afraid we are stuck with the misleading terminology.

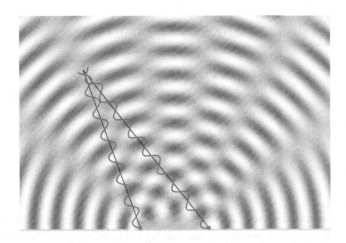

Now imagine doing the same thing with *light*. The wavelength is very much smaller, so to make the two sources Young scratched lines very close together on an opaque slide, and illuminated the two slits with a lantern (nowadays we use a laser). This time we don't see the nodal lines (though you *can*, if you blow smoke or chalk dust onto the beam), but on a distant screen (which is like the top edge of the figure) you do see a tell-tale pattern of bright and dark spots, corresponding respectively to points of constructive and destructive interference. If light had been a stream of particles, you would have seen just *two* spots – one for particles that came through the left slit, and one for particles that came through the right slit. But in fact you see 10 or 20 spots, before they fade away at the edges of the screen. From the separation of the spots and the geometry of the arrangement, you can even determine the wavelength of the light.[26]

Problem 44. Two loudspeakers, mounted 3 m apart on a wall, are driven in unison by the same amplifier, delivering a sustained note with a wavelength of 2 m. You are standing 4 m in front of one of the speakers, as shown in the figure below.

(a) How far are you from the other speaker?

(b) How many wavelengths are you from each speaker?

(c) What do you hear?

(d) If you move 1.5 m to the right (so you are the same distance from both speakers), what will you hear?[27]

---

[26] You can easily reproduce Young's double-slit experiment for yourself, if you have access to a laser. Cut the slits in a 3×5 card, using a razor blade.

[27] In practice there will be reflections off the ceiling, the furniture, etc., so this doesn't work perfectly.

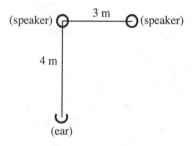

1.5.3 Standing waves

When the reflected wave gets back to your hand, it reflects again, so we now
have *two* waves going down, and one coming back. When the twice-reflected
wave reaches the tree, it generates a second returning wave, and so on. In
general, these multiply reflected waves are out of phase, and they tend to cancel
each other out – the rope jiggles a bit, but there's nothing dramatic. However: if
you shake it at just the right frequency so that the multiply reflected waves are
exactly in phase, then they all add up, making one big wave in each direction.[28]
This is an example of **resonance**: a particular frequency at which a system just
loves to oscillate.

When you have two waves of the same amplitude and frequency, propagating
in opposite directions, the result is a **standing wave**; the rope vibrates up and
down, but there is no net wave motion to the left or right.

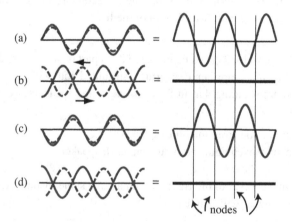

---

[28] Indeed, the waves should grow bigger and bigger without limit, as more and more reflections
pile up. In practice, friction limits the growth.

At one instant (a) the waves superimpose constructively. A moment later (b) one wave (the solid line) has moved to the right, the other (dashed) to the left – they now exactly cancel, and the rope is instantaneously flat. Later still (c) they again interfere constructively, but this time in the opposite direction. Finally (d) they again cancel. The net result (on the right, in the figure) is that the rope simply oscillates up and down. Interestingly, there occur **nodes**, a distance $\lambda/2$ apart, where the rope never moves at all.

Resonance occurs when one full "lobe," or two, or three, ... fits perfectly.

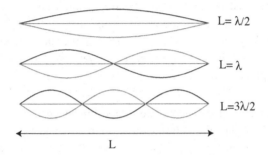

Thus

$$L = n\frac{\lambda}{2}, \quad (n = 1, 2, 3, \ldots), \tag{1.35}$$

where L is the length of the rope. In terms of frequency (remember, $\lambda f = v$, the wave speed),

$$f = \left(\frac{v}{2L}\right) n. \tag{1.36}$$

If you are shaking the rope, and you gradually increase the frequency, most of the time there will not be much response (because the multiply reflected waves are all out of phase, and tend to cancel out). But when you hit one of the resonant frequencies, suddenly the whole thing oscillates in unison, and the response is large.

This is, incidentally, the basis for all stringed instruments, including pianos and harpsichords, as well as violins and guitars. When you pluck, bow, or hammer the string, you are stimulating all frequencies, but the string responds significantly only at the resonant frequencies – the "fundamental" ($n = 1$), the "first overtone" ($n = 2$), and so on. Wind instruments are similar, only now it is standing sound waves in a pipe that create the tone.

Problem 45. The active part of a guitar string is 60 cm long. What is the wavelength of the fundamental ($n = 1$)? What is the wavelength of the "third harmonic" ($n = 3$)?

Problem 46. A violin has been tuned[29] so that the velocity of waves on the E string (33 cm long) is 435 m/s.

(a) What is the wavelength of the fundamental? What is its frequency?
(b) The vibrating string sets up sound waves in air. Their *frequency* is the same as the frequency of the waves on the string (of course), but their *wavelength* is completely different, because the speed of sound in air (340 m/s) is not the same as the speed of waves on the string. Find the wavelength of the resulting sound wave.

In this chapter we have encountered three fundamental physical entities: particles (chunks of matter), fields (mediators of forces), and waves (oscillations of a continuous medium). As we shall see, these three concepts inform all of twentieth-century physics, but in ways nobody could have anticipated.

---

[29] The speed of waves on a string is $v = \sqrt{TL/m}$, where T is the tension and m is the mass. When you tune an instrument, you are actually adjusting T, and hence v.

2

Special relativity

Classical physics,[1] some aspects of which we discussed in Chapter 1, is – for the most part – comforting to our intuitions. You probably wouldn't have come up with Newton's second law ($F = ma$) on your own (after all, nobody did before Newton), but once it is on the table it feels right. It seems consistent with our everyday experience. Classical physics refines and perfects our intuitions, but it doesn't upset them. By contrast, the four revolutions in twentieth-century physics are wildly counterintuitive; they seem to contradict everything we thought we understood – everything we took for granted about the world. That is, in part, what makes them so interesting. But it also raises a recurring question: "If this is really true, how come I never noticed it before?" I hope you will keep a skeptical eye on that subtext, as we go along.

2.1 Einstein's postulates

Einstein published his **Special Theory of Relativity** in 1905. The special theory is not an account of any particular physical phenomenon; rather, it is a description of the *arena* in which *all* phenomena occur. It is a theory of space and time themselves. As such, it takes precedence over all other theories. If you were to propose a new model of elementary particles, say, the first thing to ask would be, "Is it consistent with special relativity?" If not, you have some fast talking to do. As Kant would say, special relativity is a prolegomenon to any future physics.

Einstein based the theory on two postulates.

Postulate 1: The principle of relativity.
Postulate 2: The universal speed of light.

---

[1] Roughly speaking, "classical" physics is the subject as it stood in the year 1900.

In the following sections I'll explain these postulates. After that we'll consider their implications.

2.1.1 The principle of relativity

The principle of relativity says that the ordinary **laws of physics apply just as well in a system moving at constant velocity as they do in one at rest**. Imagine a train, traveling at constant speed down a smooth straight track. In the parlor car there is a billiard table. The principle of relativity says that the billiard game will proceed in this moving car just the same as it would if the train were parked in the station.[2] You don't have to "correct" your shots for the motion of the train. Indeed, if you pull all the curtains you will have no way of knowing whether the train is moving, since all physical processes will play out exactly as they would in a train at rest. (By contrast, if the train speeds up or slows down, or rounds a corner, or goes over a bump, you know it immediately: the balls go in weird curving trajectories, you lurch to one side, and spill coffee all over your shirt. The principle of relativity applies *only* to systems moving at *constant speed*, in a fixed direction.)

There was nothing new about the first postulate – Galileo had stated it very clearly 250 years before Einstein. One of the objections to the Copernican notion that the Earth orbits the Sun – rather than the other way around – had been that if the Earth were moving, we would surely have noticed it. When you drop something, so it was claimed, the object wouldn't fall straight down – it would "drift back" as the Earth moves out from under it. Galileo supposedly[3] climbed the mast of a moving boat and dropped a rock; it landed directly at the base of the mast, just as it would have if the boat had been at rest.[4] By the same token, Galileo argued, you cannot detect the motion of the Earth. Einstein simply appropriated Galileo's principle. If he added anything, it was the assertion that it applies to *all* physical phenomena (including, for instance, electricity, magnetism, and light), whereas Galileo would probably have limited it to classical mechanics.

An awkward implication is that there is absolutely no way of knowing whether you are at rest or not, and if that is the case, how would you know

---

[2] Einstein liked to illustrate his arguments with homey parables like this – "thought experiments", as he called them (in German, *Gedankenexperimente*).

[3] Most stories like this turn out to be historical fiction, but that doesn't diminish their explanatory power. If Galileo didn't actually do it, he could have and should have.

[4] In the reference frame of the boat, the rock was released from rest, and dropped straight down. In the reference frame of someone standing on the dock, the rock already had a forward velocity when it was released; it followed a curved (parabolic) trajectory, but since its constant forward velocity was the same as the boat's, it still landed at the base of the mast. The two descriptions are different, but they are both correct, and they arrive at the same final conclusion.

if someone else is moving at "constant velocity"? Evidently our statement of
the principle of relativity is defective, because it makes reference to something
that can never, in principle, be known. In the old books they would say the
"rest system" is the one attached to the "fixed stars," but today we know that
the stars are anything but fixed. So that's no good. How are we to break into
this logical circle? One solution is to define an **inertial reference frame** as
a system in which Newton's first law, the law of inertia, holds (objects keep
moving with constant velocity, unless acted on by some force). If you want to
know whether you are in an inertial system, grab a pile of rocks and throw them
around in various directions. If they keep going in straight lines with constant
speed, you've got yourself an inertial reference frame – and any system moving
at constant velocity with respect to you is another inertial frame. Now I can
give a more sophisticated statement of the first postulate: **the ordinary laws of
physics hold in any inertial system**.

> **Example 1.** A 3 kg lump of clay, going 4 m/s, hits a 1 kg lump
> at rest. If they stick together, what is the velocity of the resulting
> composite lump?

> (before)   (after)

Solution: Use conservation of momentum:

$$(3)(4) = (3 + 1)(v) \quad \Rightarrow \quad v = 3 \, \text{m/s}.$$

Now suppose that the collision took place on a train, which
is moving along to the right at 1 m/s. The velocities I gave you
were with respect to the *ground* (where we know conservation of
momentum holds). From the perspective of someone riding on the
train, the process looks like this:

(before)   (after)

(The 3 kg mass, for example, which is going 4 m/s with respect to
the ground, is going 3 m/s relative to the train – in general, you just
subtract 1 m/s from all ground velocities to get the corresponding
train velocities.)

Question: Is momentum conserved in this moving reference frame? Well,

$$p_i = (3)(3) + (1)(-1) = 9 - 1 = 8, \quad \text{and} \quad p_f = (4)(2) = 8.$$

Yes: momentum is also conserved on the train. The *numbers* are completely different, but the same *law of physics* prevails. The principle of relativity says this will be true for *all* processes.

Problem 1.

(a) A 5 kg cart, going 3 m/s, rear-ends an identical cart going 1 m/s. If they stick together, what is the speed of the combination, after the collision?

(b) Now examine that same process from the perspective of someone moving along at 2 m/s. How fast would *she* say the first cart is going? How fast is the second cart is going? How about the combination?

(c) Is momentum conserved, in this moving reference frame?

2.1.2 The universal speed of light

The second postulate[5] says that **light travels** (in vacuum) **at the same speed** ($c = 3 \times 10^8$ m/s) **in all inertial frames** – regardless of the motion of the source. This was Einstein's truly revolutionary proposal. It may sound innocent, but a moment's reflection reveals that it is not just radical, but downright preposterous.

Suppose you're on a train, going 60 mph, and you're feeling hungry, so you head for the dining car; say your walking speed is 4 mph.

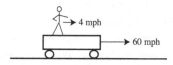

Question: how fast are you going relative to the ground? *Answer:* (obviously) 64 mph. (After an hour the train has gone 60 miles, and you have walked 4 miles up the corridor – it's a *long* train – so you've gone a total of 64 miles, and it took you 1 hour, so 64 mph.) Formally, we just *add* the two velocities:

$$v_{AC} = v_{AB} + v_{BC}. \tag{2.1}$$

---

[5] Most authors call it the "constant" speed of light, but to my ear "constant" implies "unchanging in time" ... the same today as yesterday. That is presumably *also* true of the speed of light, but it is not the issue here. What Einstein claimed is that this velocity is *universal* – the same for all observers.

In words, the velocity of A (you) relative to C (the ground) is the velocity of A relative to B (the train) plus the velocity of B relative to C. This is sometimes called **Galileo's velocity addition rule**, though it scarcely deserves so grand a title – it's really nothing but common sense.[6]

But now suppose that instead of walking down the aisle, you shine a flashlight down the aisle.

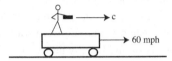

If the light beam travels at speed c relative to the train, it must, by the rule we just established, be traveling at speed $c + 60$ relative to the ground. But the second postulate says that light travels at the *same* speed (c) in all (inertial) systems – it's got to be going at c, not $c + 60$, relative to the ground. Any sensible person would conclude that Einstein has made an elementary blunder – the second postulate *cannot* be true.

Of course, Einstein wasn't stupid; he was well aware of this objection, and he traced it back to Galileo's velocity addition rule, which he replaced with

$$v_{AC} = \frac{v_{AB} + v_{BC}}{1 + (v_{AB}v_{BC}/c^2)}. \tag{2.2}$$

We call this **Einstein's velocity addition rule**. Watch how this fixes the problem: if $v_{AB} = c$ (the velocity of the light relative to the train), and $v_{BC} = v$ (the velocity of the train relative to the ground), then the velocity of light relative to the ground is

$$v_{AC} = \frac{c + v}{1 + (cv/c^2)} = \frac{c + v}{1 + (v/c)} = \frac{c(c + v)}{c[1 + (v/c)]} = \frac{c(c + v)}{(c + v)} = c.$$

The velocity of light is *automatically* c with respect to the ground, if it is c relative to the train!

That rescues the second postulate, but you're bound to wonder: if this is true, how come nobody noticed it long before Einstein? The answer is that the "correction term," $v_{AB}v_{BC}/c^2$, which adds to the 1 in the denominator, is ordinarily extremely small, because c^2 is so huge. If v_{AB} and v_{BC} are, say, 30 m/s (already a pretty good clip), then

$$\frac{v_{AB}v_{BC}}{c^2} = \frac{(30)(30)}{(300\,000\,000)(300\,000\,000)} = 0.000\,000\,000\,000\,01.$$

---

[6] Indeed, we used it almost without thinking, in Example 1.

Added to 1, this is utterly negligible. It is only when one of the velocities in question is comparable to the speed of light that there is any significant difference between Einstein's rule and Galileo's.

Still, there is something deeply disturbing about this, for Galileo's rule seems to depend on nothing more than simple common sense. How could it possibly be wrong? The answer is very subtle, and we will come to that later, but perhaps I had best tell you the essence of it now, before you lose a lot of sleep. It turns out that clocks in a moving system run slow, and meter sticks in a moving system are short. When I said you were walking at 4 mph, that was using measuring instruments on the train, whereas the train's speed (60 mph) was measured using instruments on the ground. Adding those two numbers is like adding yards to meters – they're not the same thing. Einstein's formula fixes this, in effect measuring everything relative to the ground, so that the numbers we are combining are compatible.

How did Einstein ever come up with the second postulate? Actually, he was puzzled by internal coincidences in the theory of electricity and magnetism, and only by a roundabout route arrived at the universal speed of light. But a critical experiment had already been done (though Einstein was only dimly aware of it) that quite directly suggests the universal speed of light: the **Michelson–Morley experiment** (1887). Maxwell's theory of electricity and magnetism predicted that electromagnetic waves should propagate at a speed that he calculated to be about 3×10^8 m/s. The conclusion – that light *is* an electromagnetic wave – was inescapable, but this raised an obvious question: 3×10^8 m/s *relative to what?* If we were talking about waves on a pond, the speed would be relative to the water; for sound waves, it is relative to the air. In the case of light the "medium" (whatever it is that does the waving) is not so obvious, and it was simply called **ether**.

Now, the Earth must be moving with respect to the ether – after all, it rotates once a day about the North–South axis, it orbits once a year around the Sun, the Solar System is in motion around the Galaxy, and for all I know the Galaxy itself is traveling at high speed through the cosmos. It would be a miracle if we just happened to be at rest with respect to the ether. Like a motorcycle on the open road, we should be buffeted by a strong "ether wind."[7] But which way is the wind blowing, and what is its speed? Michelson and Morley undertook to find out.

In principle the experiment is ridiculously simple; all you need is a meter stick, a stopwatch, and a flashlight. Measure the speed of light in various

---

[7] Either that, or the Earth has some sort of "windshield," and drags its local supply of ether along with it. That's actually what Michelson and Morley concluded, but subsequent experiments using starlight showed it to be false.

directions: one way, where it is swept along by the ether wind, it should go relatively fast; in the opposite direction, bucking the current, it should be relatively slow.

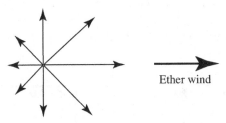

Speed of light in different directions

Of course, because light travels inconveniently fast, you can't literally do it with a stopwatch and a meter stick, so Michelson and Morley had to be more clever. The details do not concern us; what they in fact discovered is that the speed of light is the same in all directions. Evidently the speed of light is *independent* of the motion of the Earth; it is not c relative to the ether, but c *period*.[8]

Problem 2. A "moving sidewalk" at some future Los Angeles airport goes at 9/10 the speed of light. A traveler in a real hurry runs along it at 4/5 the speed of light. What is this traveler's speed relative to the ground, (a) according to Galileo; (b) according to Einstein? (Leave your answers in the form of a multiple of c. Note that (a) is greater than c, but (b) is less than c.)

Problem 3. As the outlaws escape in their get-away car, which goes $\frac{3}{4}c$, a cop fires a bullet from the pursuit car, which only goes $\frac{1}{2}c$. The muzzle velocity of the bullet is $\frac{1}{3}c$. Do the bad guys get away, (a) according to classical physics; (b) according to Special Relativity?

2.2 Implications

In the following sections we shall explore the implications of Einstein's postulates. For a while it is going to seem as though everything you took to be "obvious" about the world is incorrect. Do not be too alarmed: really, there are only three or four fundamental surprises – after that everything begins to fall into place.

---

[8] This being the case, the whole notion of "ether" loses its appeal, and the term has died out. But it does afford a cute way to interpret the two postulates: (1) there is no absolute rest, (2) there is no ether.

2.2.1 The relativity of simultaneity

We are back on the train, and someone has wired up a lightbulb, hanging from the ceiling at the very center of the car. When you switch it on, light goes out in all directions. Detectors (*a* and *b*) have been installed at the front and back ends of the car – maybe a buzzer sounds when the light hits the detector. *Question:* Which buzzer rings first?

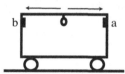 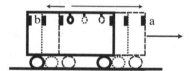

Well, from the perspective of an observer on the train, they both ring at the same time (simultaneously), because the light has just as far to travel in both directions. But from the perspective of an observer on the ground, the train itself moves a bit during the process, so the light going to the back end has a *shorter* distance to go, and the light heading to the front has a *longer* way to go. According to the second postulate, remember, light travels at the same speed in both frames and in all directions. It follows that in the ground frame of reference buzzer *b* sounds before buzzer *a*. *Conclusion:* **Two events that are simultaneous in one inertial frame are not, in general, simultaneous in another.** Naturally, the train has to be going awfully fast before the discrepancy becomes detectable – that's why we don't notice it all the time.

Of course, it is always possible to be *mistaken* about simultaneity. If you happen to be standing at the rear end of the car, you may *hear* buzzer *b* before you hear buzzer *a*, just because the sound gets to you more quickly. It doesn't take an Einstein to figure that out – *obviously* you need to correct for the time it took the news to reach you, whether it comes by sound, by light, or by Federal Express. We use the word **observation** (by extension **observe** and **observer**) to denote what you get *after* you have made the necessary correction for the time it took the data to reach you. What you observe, then, is not at all the same as what you see or hear.

For example, when you look up at the night sky, the light you see left the Moon 1.3 seconds ago, it left Jupiter around 40 minutes ago, it left the star Sirius 8.6 years ago, and it left the Andromeda galaxy 2.5 million years ago. You're not seeing distant objects as they are in 2013, but as they were at various times in the remote past. If you want to know what's happening up there *right now*, you're going to have to wait for the news to get here.[9] To *observe* the

---

[9] Of course, in most cases the action is much closer to home, and the wait is very short. I'm just trying to emphasize the point of principle.

night sky (in the technical sense), you would assemble all that information as a reconstruction, after the fact. An observation can be made only when all the data are in.[10]

Relativity has to do with what you *observe*. We are talking about real effects, not accidental appearances. Einstein liked to say that all the conceptual difficulties of special relativity derive ultimately from the relativity of simultaneity. It is subtle and counterintuitive, so be on guard!

Problem 4. Synchronized clocks are stationed at regular intervals, a million kilometers apart, along a straight line. When your clock reads 12 noon, (a) what time do you see on the nineteenth clock down the line? (b) What time do you observe on that clock?

2.2.2 Time dilation

Now let's move one of the detectors down to a point on the floor directly beneath the lightbulb. Call the distance from the bulb to the detector h. *Question:* How long does it take the light signal to reach the detector?

According to an observer on the train, the light goes a distance h, at speed c, so $ct_t = h$, or

$$t_t = \frac{h}{c} \tag{2.3}$$

(t_t is the time it takes as measured on the train).

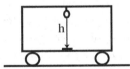

But from the perspective of an observer on the ground, the train moves a distance vt_g, in the time t_g it takes the signal to reach the detector, so that *same light ray* travels along the hypotenuse of a right triangle to reach the detector.

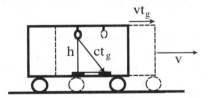

---

[10] You could, in principle, station assistants all over the Universe, to collect data right at the scene, with no time delay to worry about. But you're still going to have to wait for their reports to arrive before you can make your observation.

According to the Pythagorean theorem,

$$(ct_g)^2 = (vt_g)^2 + h^2, \quad \text{or} \quad c^2 t_g^2 \left(1 - \frac{v^2}{c^2}\right) = h^2,$$

so

$$t_g = \frac{h}{c} \frac{1}{\sqrt{1 - v^2/c^2}}. \tag{2.4}$$

The factor $1/\sqrt{1 - v^2/c^2}$ is ubiquitous in relativity; we call it γ (the Greek letter gamma):

$$\gamma = \frac{1}{\sqrt{1 - v^2/c^2}}. \tag{2.5}$$

Notice that γ is a number *greater than or equal to* 1 – and equal only in the case $v = 0$, when the train is not moving at all.[11] Comparing Eqs. (2.3) and (2.4),

$$t_g = \gamma t_t. \tag{2.6}$$

The time it takes, as measured on the ground, is *greater* than the time it takes as measured in the train! If the ground clock ticks off 10 minutes, the train clock only ticks off 7 minutes. Evidently the train clock is running slow.

Moving clocks run slow (by a factor of γ).

This is called "**time dilation**."

> **Example 2.** Time dilation is perhaps the easiest relativistic effect to demonstrate in the laboratory. It so happens that most elementary particles spontaneously disintegrate, after a characteristic "lifetime." A muon (pronounced "m'you'-on") at rest, for example, lasts $2 \times 10^{-6} = 0.000\,002$ s. Such a particle constitutes a tiny clock (it only ticks once, but never mind). Now suppose a muon is traveling at high speed through the laboratory. Moving clocks run slow, so it actually lasts *longer* than it would have had it been at rest. This is observed all the time at places like the Stanford Linear Accelerator Center (SLAC).
>
> Say, for example, that a muon is going at $v = 3/5\,c$. Then
>
> $$\gamma = \frac{1}{\sqrt{1 - (3/5c)^2/c^2}} = \frac{1}{\sqrt{1 - (9/25)}} = \frac{1}{\sqrt{16/25}} = \frac{1}{(4/5)} = \frac{5}{4}.$$

[11] For v greater than c, γ becomes imaginary (the square root of a negative number). This is our first indication that it is impossible (in relativity) for any object to travel faster than light.

This muon will last

$$t_g = \frac{5}{4}(2 \times 10^{-6}) = 2.5 \times 10^{-6} \, \text{s}.$$

In this time it will go a distance

$$vt_g = \frac{3}{5}(3 \times 10^8)(2.5 \times 10^{-6}) = 450 \, \text{m}.$$

Absent time dilation, it would have gone $(3/5)(3 \times 10^8)(2 \times 10^{-6}) = 360 \, \text{m}.$

Problem 5. It takes Johnny 7 minutes to eat a hot dog, according to his own watch. If Johnny is on a high-speed train, going at 90% the speed of light, how long does it take him to eat that hot dog according to observers on the ground, who watch him through the dining-car window?

Problem 6. The star nearest Earth, Alpha Centauri, is 4.3 light years[12] away. If a spaceship makes the (one-way) trip at a constant velocity of $0.95 \, c$, how long does it take, according to an Earth clock? How long does it take according to a clock on the spaceship?

Problem 7. Muons are produced high in the atmosphere (say, 10 km up) by the action of cosmic rays. Their velocities are typically close to the speed of light – say, $0.999 \, c$. We know that the average lifetime of a muon at rest is $2 \times 10^{-6} \, \text{s}$. Would such a muon ever reach ground level, (a) according to classical laws; (b) according to special relativity? [*Hint:* calculate how far the muon would travel, in each case.]

2.2.3 Lorentz contraction

One last thought experiment. This time we mount the lightbulb on the back wall of the boxcar, a mirror on the front wall, and the detector on the back wall.[13]

---

[12] A light year is the distance light travels in one year:

$$(3 \times 10^8)(60)(60)(24)(365) = 9.46 \times 10^{15} \, \text{m}.$$

[13] Ignore the slight tilt of the light rays in the figure – I'm just trying to separate the forward beam from the return beam. Also ignore the size of the light bulb and the detector – I enlarged them for clarity.

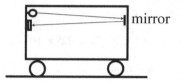

Question: How long does it take the light signal to go down and back? Let's say the length of the car is L_t (as measured on the train). To an observer on the train, the light signal has to go a total distance $2L_t$, and it travels (as always) at the speed c, so $ct_t = 2L_t$, or

$$t_t = \frac{2L_t}{c}. \tag{2.7}$$

But from the perspective of an observer on the ground, the light has to travel somewhat farther on the way down (because the train has meanwhile moved a bit), and somewhat less far on the return trip.

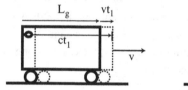 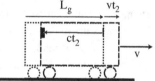

Indeed, if t_1 is the forward time,

$$ct_1 = L_g + vt_1, \quad \text{or} \quad (c - v)t_1 = L_g, \quad \text{so} \quad t_1 = \frac{L_g}{c - v}$$

(where L_g is the length of the car as measured on the ground). Likewise, if t_2 is the return time,

$$ct_2 = L_g - vt_2, \quad \text{or} \quad (c + v)t_2 = L_g, \quad \text{so} \quad t_2 = \frac{L_g}{c + v}.$$

The total time for the round trip, then, is

$$t_g = t_1 + t_2 = L_g \left[\frac{1}{c - v} + \frac{1}{c + v} \right] = L_g \left[\frac{c + v}{(c + v)(c - v)} + \frac{c - v}{(c + v)(c - v)} \right]$$

$$= L_g \frac{c + v + c - v}{c^2 - v^2} = L_g \frac{2c}{c^2(1 - v^2/c^2)} = \frac{2L_g}{c} \gamma^2. \tag{2.8}$$

But by time dilation we know that $t_g = \gamma t_t$, so, comparing Eqs. (2.7) and (2.8),

$$t_g = \frac{2L_g}{c} \gamma^2 = \gamma t_t = \gamma \frac{2L_t}{c},$$

or

$$L_g = \frac{1}{\gamma}L_t.$$

The box car is *shorter*, as measured by an observer on the ground, than it is when measured on the train! This is called **Lorentz contraction**.

Moving objects are short (by a factor of γ).

Please note that I did *not* say moving objects *appear* short, or *seem* short, or *look* short; we're not talking about appearances, here – they simply *are* short, as compared with their rest length. The manufacturer certifies that the boxcar is 50 ft long (it was at rest when they made it), and an observer riding along in the boxcar confirms that figure (the boxcar is at rest with respect to her), but an observer standing on the ground, who watches the train whiz by, measures its length and finds it to be only 40 ft. If the train now comes to a stop, and the guy on the ground measures it again, he'll get 50 ft, because the boxcar is now at rest with respect to him. Astounding!

> **Example 3.** Lorentz contraction is not so easy to demonstrate in the laboratory, for the simple reason that if something is big enough to have a measurable length, it is going to be very hard to get it going anywhere near the speed of light. But in principle, if I could get a meter stick, say, going at 3/5 c (remember, that means $\gamma = 5/4$), it would only be 80 cm long, as it flew by.

Lorentz contraction only applies to lengths *along the direction of motion*; there is no shortening of dimensions perpendicular to this. A square box, for example, becomes rectangular.

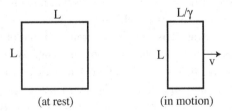

(at rest)   (in motion)

> **Problem 8.** A high-speed train leaves San Francisco at 12 noon. When it pulls into Portland, the station clock reads 5 pm. If the train has been going at 0.8 c (don't ask me how it took 5 hours to make the trip at that speed – that's just the way Amtrak operates), what time does the engineer's watch

(which he set by the station clock in San Francisco at the time of departure) read, upon arrival? If the train's length when stopped in the station is 250 m, how long was it according to an observer in Salem, who watched it fly past?

Problem 9. A Lincoln Continental is twice as long as a VW Beetle, when they are at rest. As the Continental overtakes the VW, going through a speed trap, a (stationary) policeman observes that they both have the same length. The VW is going at half the speed of light. How fast is the Lincoln going? (Leave your answer as a multiple of c.)

2.3 Paradoxes

You've probably had this experience: You're sitting on a subway, waiting to leave the station. Looking out the window, you find yourself looking *in* the window of a train on the other track, which is also stopped in the station. Finally your train starts to move. As you pass the end of the other train, the platform comes into view ... and you suddenly realize that it is they, not you, who have left the station – you're still sitting there.

This is the principle of relativity (Einstein's first postulate) in action. The other train is moving relative to you; you are moving relative to it. Which one is "really" moving? The question has no answer: all motion is relative. That's the essence of the principle of relativity.

"Nonsense," you say. "Clearly *he* is the one who is moving – I'm still sitting in the station!" Well, OK: he is moving *relative to the platform* (and hence to the Earth below), and I grant you, that's the most relevant reference frame in this context. But the principle of relativity says that his reference frame (the one in which he is at rest) is just as valid as yours.

"Nonsense," you repeat. "His wheels are turning, his motor is running, he is consuming fuel – obviously he is the one who is moving." Well, all these things have to do with the fact that trains ride on rails, that are attached to the Earth; they confirm that he is moving relative to the tracks. But suppose instead it was a boat floating down a river – then no motor would be required for it to move (relative to the bank), and a motor *would* be required to keep it stationary. So the motor doesn't tell you whether something is moving. Or imagine two rockets passing each other in empty space. In that case (with no Earth to confuse the issue) it would be impossible to say which is moving – *each* is moving, relative to the other, and each is at rest, in its own reference frame.

A reference frame is an abstract entity – it doesn't have wheels or motors; it is equipped with clocks and meter sticks (or rather, with time and distance), and it accommodates observers to record events (specifically, the time and place at which they occur). Physical objects, like trains and rockets and people, can be at rest or in motion with respect to a given reference frame, but there is no such thing as absolute rest or absolute motion.

Much of special relativity consists of examining some process from the perspective of two different reference frames, and comparing the observations made. The famous "paradoxes" of relativity are delightful examples. They are not *really* paradoxes, of course – no logical inconsistency is involved. But it can certainly *seem* like it, when you first encounter them, so fasten your seat belt!

2.3.1 The paradox of Lorentz contraction

I come running into the room, carrying a meter stick.

You (sitting at your desk) call out, "Hey, buddy – your meter stick is only 80 cm long. Better get a new one!" That's Lorentz contraction. But from *my* perspective it is *you* who are moving; I shout back, "On the contrary. My meter stick is fine – in fact, *yours* is short!" You say my meter stick is short; I say yours is short ... who is right? *Answer:* We *both* are! How can that be? Surely, if my meter stick is too short I should have concluded that yours is too *long*.

To resolve this paradox we have to examine closely what it *means* to measure the length of something. Say I want to know the length of this dark line.

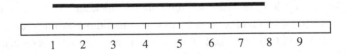

I might hold a ruler next to it, note the number at the right end, and subtract the number at the left end – in this instance $8 - 1 = 7$ cm. (If I'm really clever I will line up the left end of the ruler with the left end of the line; then I only need to read *one* number.)

OK, but what if the thing I'm measuring is *moving*? Maybe it's a centipede, crawling across the page. Same story: hold up the ruler and record the two end numbers. Only this time (obviously) I must take care to read the two numbers *at the same time*. It wouldn't do to record the position of the head, and then wait a minute before recording the position of the tail, because the centipede would meanwhile have moved a bit, so of course I'd get the wrong answer. For a valid measurement, you must read the two ends *simultaneously*.

And that's the whole problem. What's simultaneous to *me* is not simultaneous to *you* (that's the relativity of simultaneity). Watching you measure my meter stick, I see you make the most elementary blunder in the book: you recorded the front end and then *waited* before recording the back end. That's how, even though (from my perspective) you were using a meter stick that was itself too short, you managed to conclude that *mine* is too short. Of course, you have exactly the same objection to my measurement of your meter stick. Each of us did exactly the right thing, within our own reference frame, but each of us thinks the other made a foolish error – and it's all due to the relativity of simultaneity.

2.3.2 The paradox of time dilation

Now I come running into the room carrying a clock.

You call out, "Hey, buddy – your clock is running slow. Better get a new one!" That's time dilation. But from my point of view it is you who are moving; I shout back, "On the contrary. My clock is fine; in fact, *yours* is running slow!" Who's right? We both are! How can that be?

Well, let's watch closely how you measured the rate of my clock. When I first entered the room, Michael (who was sitting close to the door) recorded the time on my clock and on his, and when I got to the far side of the room, Nora (sitting by the window) recorded the time on my clock and the time on hers. Subtracting Michael's number from Nora's, you found that while your clocks ticked off 15 seconds, mine ticked off only 10 seconds, and you concluded that my clock was running slow. But notice that you compared my one clock with two *different* clocks (Mike's and Nora's). "So what?" (I hear you say), "We carefully synchronized them in advance, so it doesn't matter." Ah, but what does it mean to synchronize your clocks? It means that you set them both to read 12 noon (say) at the same time – *simultaneously*. Because of the relativity of simultaneity, from *my* perspective your clocks weren't synchronized in the first place! So it was absurd for you to subtract Michael's time from Nora's – Nora's was 10 seconds ahead to begin with. That's how even though both Mike's and Nora's clocks were in fact running slow (according to me), you managed to conclude that *my* clock was running slow. Of course, you make the same objection to my measurement of your clocks. Both of us did exactly the right thing, within our own reference frame, but each of us thinks the other made a silly blunder: using clocks that weren't synchronized.

2.3.3 The barn and ladder paradox

A farmer has a ladder that is too long to fit in his barn.

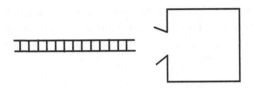

He has been reading some relativity, and it occurs to him that if his son runs with the ladder at high speed, it will Lorentz contract, and in this way he will be able to get it into the barn.

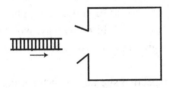

But his son, who has also been reading about relativity, counters that from his perspective it will be the *barn* that contracts, and the fit will be even worse.

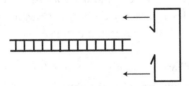

Who is right? Will the ladder fit in the barn, or won't it?

Once again, they are both right. What does it *mean* to say that the ladder "is in the barn"? I guess it means that both ends of the ladder are in the barn at the same time (simultaneously). Really, there are two relevant events here: (1) the back end of the ladder makes it in the barn door, and (2) the front end of the ladder hits the far wall of the barn. The farmer says (1) happens before (2) (and hence the ladder is in the barn); his son says (2) happens before (1) (so the ladder is never completely inside the barn). Contradiction? No – just a difference in perspective.

Ultimately, of course, the farmer is going to want the ladder *at rest* in the barn, and as it slows down it must return to its uncontracted length: it stretches out, and something's gotta give. Either the ladder smashes through the far wall, or the ladder breaks. In either case, the farmer is not likely to be pleased with the outcome.

Problem 10. A farmer has a 30 ft ladder, but his shed is only 20 ft long. He instructs his son to run with the ladder at $4/5\,c$, hoping that Lorentz contraction will enable him to fit the ladder into the shed. (a) According to the farmer, how long is the moving ladder? Does it make it into the shed? (b) From the son's perspective, how long is the shed? Does the ladder fit? (c) Who's right – does the ladder make it into the shed, or doesn't it?

2.3.4 The twin paradox

Paul and Mary are twins. On their twentieth birthday, Mary gets on a high-speed rocket ship and travels to a distant star. The trip takes five years, by her watch. Upon arrival, she immediately turns around and heads home, at the same speed. At their reunion, they are surprised to find that Paul has aged much more than Mary; he is celebrating his fiftieth birthday, say, while she is celebrating her thirtieth – and he has the grey hair and wrinkles to show for it. Why does the traveling twin age less?

Well, from Paul's perspective she has been traveling at high speed, so her clock has been running slow (it doesn't matter whether she is outbound or inbound, by the way – the dilation factor γ only depends on v^2, so it's the same whether v is positive or negative). And it's not just her clock that runs slow: time *itself* runs slow, for her. Her heart-beat is slow, her metabolism is slow, her thought processes are slow, everything. So she is younger, when they get back together.[14]

The "twin paradox" arises when you try to tell the story from Mary's perspective. Paul takes off on "rocket Earth," and from her point of view it is *his* clock that runs slow. Shouldn't *he* be the younger one, at their reunion? No! In this case Paul is right, and Mary is wrong. The situation is not symmetrical, because at the moment Mary stepped from the outbound rocket to the inbound rocket she underwent acceleration (she was not in an inertial reference frame) – and she is well aware of it: she felt the lurch, and she spilled her coffee. At that moment, all bets are off; the principle of relativity does not apply, and she has

---

[14] This is really true. If you desperately want to know what will be happening on Earth 10 000 years from now, in principle you could arrange it, by taking a long trip at high speed. But it's not really a fountain of youth, because you do not live any *longer* (by your own clock), you just do it *slower*.

no right to assume simple time dilation. Notice that Paul underwent no such acceleration – he never spilled his coffee, he was always in an inertial system, and his analysis is correct.

That resolves the "paradox." But it does seem strange that Mary's brief period of acceleration should have such striking consequences. I like to imagine that she carries a magic telescope, through which she can see everything that is happening *right now.* (Of course, no such instrument is possible – you have to wait for the news to reach you, and no information can travel faster than the speed of light. But never mind – it makes it easier to tell the story.) During her travels, Mary watches Paul's clock (back on Earth) through her magic telescope. On the way out, she notes that his clock is running slow (from her perspective), and he ages less than she does. On the way back, his clock is again running slow, and again he ages less. But at the moment she steps from the outbound rocket to the inbound rocket (when she is briefly in a noninertial system), she is astonished to see Paul's clock advance through many years – his hair turns grey and his muscles flabby (but in some compensation he raises a family and makes a lot of money in real estate) – all this during the two or three minutes (of *her* time) it took to transfer from one rocket to the other. That's how, even though he aged less than she did on the way out, and again on the way back, at their reunion he is the one who is older. Did anything peculiar actually happen to him, during this strange interval? Of course not – he simply lived his life, as usual. What *did* change radically (and this is very subtle, so think about it) is Mary's notion of what "right now, back home" (what she sees through the magic telescope) *means.* Simultaneity is completely different for the inbound rocket frame.

Problem 11. On your twenty-first birthday you depart for a nearby star, on the express rocket, which travels at 3/5 the speed of light. At your destination you immediately catch the return flight (which also goes at $3/5\,c$), arriving home just in time to celebrate your twenty-fifth birthday. How old is your twin brother (who stayed at home)?

2.4 Relativistic mechanics

After working out the "geometrical" implications of his two postulates (the relativity of simultaneity, time dilation, Lorentz contraction, and the velocity addition formula) Einstein turned his attention to the laws of physics themselves. Some of them had to be modified to accommodate the theory of relativity – most importantly, the laws of conservation of momentum and conservation of energy.

2.4.1 Mass and momentum

Take another look at Example 1, where we examined a collision between two lumps of clay – a 3 kg lump going at 4 m/s hit a 1 kg lump at rest, and they stuck together. From conservation of mass we knew that the resulting composite lump was 4 kg, and from conservation of momentum we found that it was going at 4 m/s. We then studied the same collision from the perspective of a moving observer, and discovered that momentum was *also* conserved in that reference frame (mass, too, but we simply *assumed* that). This was supposed to illustrate the principle of relativity: the same laws (conservation of mass and conservation of momentum) hold in both reference frames.

However, to calculate the velocities in the moving frame I (implicitly) used Galileo's velocity addition rule (Eq. (2.1)). We now know that there should have been an extra factor in the denominator (Eq. (2.2)), and this spoils the whole argument. If momentum is conserved in the first system, it is *not* conserved in the second! The conservation laws, as we took them from classical mechanics, are not consistent with special relativity.

To fix this problem, Einstein had to assume that the *mass* of an object depends on its velocity:

$$m_r = \gamma m, \quad \text{where (as always)} \quad \gamma = \frac{1}{\sqrt{1 - v^2/c^2}}. \qquad (2.9)$$

Here m_r is the **relativistic mass**, and m is the **rest mass** (the mass it has when it's not moving). Remember, γ is a number larger than 1; the faster it is going the greater its (relativistic) mass.

What does it *mean* to say that the mass of a moving object is greater than when it's at rest? Well, mass is the measure of *inertia*, remember; it tells you how hard it is to accelerate the object. You would need a much larger force to accelerate something from 290 000 000 m/s to 290 000 001 m/s in (say) 1 second, than from 10 m/s to 11 m/s. In fact, it would take an infinite force to accelerate it all the way up to the speed of light.

Meanwhile, Einstein defined **relativistic momentum** as (relativistic) mass times velocity:

$$p_r = m_r v = \gamma m v = \frac{mv}{\sqrt{1 - v^2/c^2}}. \qquad (2.10)$$

This cures the problem. If *relativistic* mass and *relativistic* momentum are conserved in one inertial frame, they are automatically conserved in any other inertial frame. Of course, this doesn't prove that they *are* conserved – that's a matter for experiments to decide – but at least it makes for *possible* conservation laws (laws consistent with the postulates of relativity). And as it turns out the experiments have richly confirmed the following:

Relativistic mass and relativistic momentum are conserved

(the sum of the initial values is equal to the sum of the final values).

> **Example 4.** That 3 kg lump[15] of clay, going now at 4/5 c, hits the 1 kg lump at rest, and they stick together. Find the (rest) mass and velocity of the resulting composite lump.

(before)   (after)

Solution:

$$\frac{1}{\sqrt{1-(4/5)^2}} = \frac{1}{\sqrt{1-(16/25)}} = \frac{1}{\sqrt{9/25}} = \frac{1}{3/5} = \frac{5}{3},$$

so the relativistic mass of the 3 kg lump is $m_r = (5/3) \times 3 = 5$ kg, and its relativistic momentum is $p_r = (5) \times (4/5)c = 4c = 1.2 \times 10^9$ kg m/s. Before the collision the total relativistic mass was $(5 + 1) = 6$ kg, so conservation of relativistic mass and momentum says $M_r = 6$, and $M_r v = 4c$. Thus $6v = 4c$, so $v = 2/3\,c$, and hence

$$\gamma = \frac{1}{\sqrt{1-(2/3)^2}} = \frac{1}{\sqrt{1-4/9}} = \frac{1}{\sqrt{5/9}} = \frac{1}{\sqrt{5}/3} = \frac{3}{\sqrt{5}}.$$

Therefore, the final rest mass is

$$M = \frac{M_r}{\gamma} = \frac{(6)\sqrt{5}}{3} = 2\sqrt{5} = 4.47 \text{ kg}.$$

Notice that *rest* mass is *not* conserved (in Example 4, (3+1)=4, not 4.47)! As we'll see in Chapter 4, most of the interesting processes in particle physics would be impossible, if rest mass were conserved. For instance, the neutron decays into a proton, an electron, and a neutrino,

$$n \to p + e + \nu,$$

but the (rest) masses on the right don't add up to the mass of the neutron. Their *relativistic* masses are slightly larger, and these *do* add up to the mass of the neutron. Relativistic mass is conserved, but rest mass is not.

---

[15] Unadorned, "mass" always means "rest mass."

Problem 12. At what velocity is the relativistic mass of an object equal to three times its rest mass?

Problem 13. Particle A, of mass m, is at rest when it decays into two identical particles B, each of (rest) mass $(2/5)m$: $A \to B + B$. The Bs fly off in opposite directions. What is the speed of each B? [Leave your answer as a multiple of c.]

Problem 14. A 1000 kg Prius is going at three-fifths the speed of light (this model is not yet on the market), when it hits a 400 kg moose standing in the middle of an icy road. Luckily, the moose is unhurt; it lands on the hood of the Prius, and the two slide off as a unit. (a) What is the relativistic mass of the combination? (b) What is its rest mass? (c) How fast is it going?

2.4.2 Energy

Meanwhile, the **relativistic energy** of an object of (relativistic) mass m_r is

$$E_r = m_r c^2 = \gamma m c^2. \tag{2.11}$$

It is conserved, just as relativistic mass is. Indeed, since E_r is m_r multiplied by a constant (c^2), it's a matter of taste whether you choose to work with the one or the other.[16] Nowadays, most physicists prefer relativistic energy, and the whole notion of relativistic mass has largely died out.

Notice that *even when it is not moving* an object still has energy – we call it **rest energy**:

$$R = mc^2 \quad \text{(rest energy)}. \tag{2.12}$$

The remainder is energy of motion – **relativistic kinetic energy**:

$$T_r = E_r - R = (\gamma - 1)mc^2 \quad \text{(relativistic kinetic energy)}. \tag{2.13}$$

This is *not* the same as classical kinetic energy ($\frac{1}{2}mv^2$), or even (as you might have guessed) $\frac{1}{2}m_r v^2$, though it is close when the velocity is much less than the speed of light.

[16] If $A = B$, then $Ac^2 = Bc^2$, so if relativistic mass is conserved, then relativistic energy is conserved. It's a little like deciding whether to balance your checkbook in dollars or in cents – it doesn't matter, as long as you are consistent.

Example 5. A lump of clay of (rest) mass 10 kg, going at 12/13 c, collides head-on with an identical lump, going the opposite direction at the same speed. Find the relativistic mass of each lump, and the mass of the resulting composite lump.

(before) (after)

Solution:

$$\gamma = \frac{1}{\sqrt{1-(12/13)^2}} = \frac{1}{\sqrt{1-(144/169)}} = \frac{1}{\sqrt{25/169}} = \frac{1}{5/13} = \frac{13}{5},$$

so the relativistic mass of each lump is $m_r = \frac{13}{5}\, 10 = 26$ kg. Evidently the relativistic mass of the final lump is 52 kg – and since it is at rest (obviously – but if you want to cite a law of physics, it's conservation of momentum), this is also its rest mass.

Once again, rest mass was not conserved: the total was 20 kg to begin with, but 52 kg at the end! Where did the extra mass come from? Here it pays to think in terms of energy. In the beginning there was a lot of kinetic energy; by the end there is none. Kinetic energy was converted into rest energy, and hence the mass increased.[17]

Rest energies can be absolutely gigantic, because of the factor c^2. Here's an example.

Example 6. The mass of a penny is 2.5 grams. What is its rest energy?

Solution:

$$R = mc^2 = (2.5 \times 10^{-3})(3 \times 10^8)^2 = 2.25 \times 10^{14}\,\text{J}.$$

A joule (J) is a kg m^2/s^2, and a kilowatt-hour is 3.6×10^6 J, so

$$R = \frac{2.25 \times 10^{14}}{3.6 \times 10^6} = 6.25 \times 10^7\,\text{kwh} = 62\,500\,000\,\text{kwh}.$$

---

[17] Newspapers like to call this the "conversion of energy into mass," and the reverse process (as occurs most dramatically in an atomic bomb) "conversion of mass into energy." This is bad language – you can't convert energy into mass any more than you can convert meters into kilograms – they are categorically different things. What you *can* do is convert kinetic energy into rest energy. But I suppose that wouldn't sell newspapers.

At 5 cents a kilowatt-hour, that penny is worth 3 million dollars!
(That is, if you could figure out a way to convert all its rest energy
into electricity.)

Question: How much energy would it take to get an object of (rest) mass
m going at the speed of light? Well, $E_r = \gamma mc^2$, and $\gamma = 1/\sqrt{1 - v^2/c^2}$. But
if $v = c$, then $\gamma = 1/0 = $ infinity! It would take an *infinite* amount of energy.
Don't even think about it; it's not going to happen. Nothing with (rest) mass
can travel at the speed of light.

Problem 15. What is the velocity of a particle that has a kinetic energy
equal to its rest energy?

Problem 16. A lump of clay whose rest mass is 2 kg is traveling at $3/5\,c$.
Find (a) its relativistic mass (in kg), (b) its momentum (in kg m/s), (c) its
total energy (in J), (d) its rest energy (in J), (e) its kinetic energy (in J).
(f) For comparison, compute its *classical* (i.e. pre-Einstein) kinetic energy.

Problem 17. The lump of clay in the previous problem now collides head-
on with an identical lump traveling in the opposite direction at the same
speed, and they stick together. (a) What is the velocity of the final composite
lump? (That's obvious, in this case, but cite a law to justify your answer.)
(b) Find the total energy of the composite lump. (c) What is the rest mass
of the composite lump? Where did the "extra" mass come from?

Problem 18. A 9 kg lump of clay going at $4/5\,c$ hits a 5 kg lump at rest
(NB: these are the *rest* masses). They stick together, and fly off as a single
composite lump. (a) What was the original relativistic mass of the 9 kg
lump? (b) What is the relativistic mass of the composite lump? (c) What
was the original momentum of the 9 kg lump? (Leave your answers to (c)
and (d) in terms of c.) (d) What is the velocity of the composite lump? (e)
What is the rest mass of the composite lump?

Problem 19. A golf ball (rest mass 46 grams) is going at one-tenth the speed
of light. Calculate its kinetic energy (in joules) (a) according to Newton
(use $T = \frac{1}{2}mv^2$), and (b) according to Einstein (use $T_r = (\gamma - 1)mc^2$, and
calculate $(\gamma - 1)$ to at least three significant digits). [The point of this

problem is to check that for speeds well below c the two formulas give practically the same answer.]

Problem 20. Show that

$$E_r^2 - p_r^2 c^2 = m^2 c^4. \tag{2.14}$$

[*Hint:* Use Eqs. (2.10), and (2.11).]

Problem 21. Uranium-238 spontaneously disintegrates into thorium-234 plus helium-4:[18]

$$^{238}\text{U} \rightarrow {}^{234}\text{Th} + {}^4\text{He}.$$

The rest masses of these atoms are: ^{238}U: 238.050 79 u; ^{234}Th: 234.043 63 u; ^{4}He: 4.002 60 u. (The **atomic mass unit** u is 1.66×10^{-27} kg.)

(a) How much rest mass is lost in this decay? (First give your answer in u; then convert to kg.)
(b) How much kinetic energy is created in each disintegration? (That's the same as the rest energy lost.)
(c) How many disintegrations would it take to generate 1 kwh of electricity?
(d) How many kwh could you get altogether if you started with 1 gram of ^{238}U? (Actually, the half-life of ^{238}U is 4.5×10^9 years, which is kind of a long time to wait – but there are ways to speed up the process.)

2.4.3 Massless particles

In classical mechanics there is no such thing as a massless particle (a particle of zero mass). After all, what would its momentum be? $p = mv = 0$. What would its kinetic energy be? $T = \frac{1}{2}mv^2 = 0$. You couldn't exert a force on it; Newton's second law says $F = ma = 0$. And therefore (by Newton's third law) it couldn't exert a force on anything else. It's a cipher. But in relativity, there is a loophole worthy of a congressman:

$$p_r = \frac{mv}{\sqrt{1 - v^2/c^2}}, \quad E_r = \frac{mc^2}{\sqrt{1 - v^2/c^2}}.$$

Yes, the numerators are zero, but what if the denominator is *also* zero? Then we get 0/0, which is ambiguous – it could be anything. Now, the denominator

---

[18] Don't worry about what the numbers mean – we'll come to that in Chapter 4.

is zero if $v = c$. So in relativity it is just conceivable that there *could* be massless particles, provided that they always travel at the speed of light! In view of Eq. (2.14), the energy and momentum of a massless particle are related by

$$E_r = p_r c. \tag{2.15}$$

I would regard this as a joke, were it not for the fact that there *are* massless particles (photons – "particles" of light); they *do* travel at the speed of light, and their energy and momentum *are* related by Eq. (2.15). So, like it or not, we have to take the loophole seriously. But this raises a perplexing question: What is the *difference* between a photon (say) with a lot of energy, and one with not so much energy? "Well," you say, "the one with more energy is heavier." No! They're both massless. "Oh, then I guess the more energetic one is going faster." Nope, they're both going at the same speed (c). So what *is* the difference? Relativity offers no answer to this question, but, curiously, quantum mechanics *does*: the high-energy photon is *blue*, and the low-energy photon is *red!*

Problem 22. A typical blue photon has a (relativistic) energy of 4.30×10^{-19} J. (a) What is its (relativistic) momentum? (b) What is its (rest) mass? (c) What is its speed? (d) Using $p_r = m_r v$, what is its relativistic mass?

Problem 23. A neutral pi meson (at rest) decays into two photons:[19] $\pi^0 \to \gamma + \gamma$. The (rest) mass of a π^0 is 2.40×10^{-28} kg. (a) What is the energy of each photon? (b) What is the momentum of each photon?

2.5 The structure of spacetime

Every physical phenomenon involves a collection of **events**. An event, in the technical sense, is something that happens at a particular place, at a particular time. The explosion of a firecracker is an event; your trip across the country is not. We represent the coordinates of an event on a spacetime graph (or **Minkowski diagram**).

---

[19] This is, by the way, a rare example of 100% conversion of rest energy into kinetic energy.

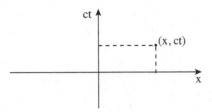

By convention, the position (x) is plotted horizontally, and time (t – or rather, ct) vertically.

On a Minkowski diagram, any object traces out a path (called its **world line**), as time progresses.

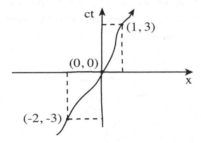

This particular object was at $x = 0$, at time $t = 0$; a little earlier it was at $x = -2$, and a little later it was at $x = 1$. A vertical line would represent a particle at rest (it's always at the same place); a straight line (but not vertical) would represent a particle moving with constant velocity – the steeper the slope, the slower it is going. A 45° line would represent something traveling at the speed of light ($\Delta x = c\Delta t$).

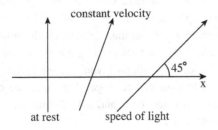

Because no object can travel faster than light, world lines can never have slopes less than 45°. At any given moment, the future of the world line is confined to a "wedge," bounded by the two 45° lines.

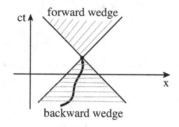

That's the locus of all points in spacetime the object could possibly get to. Similarly, the backward wedge is the locus of all points from which it could conceivably have come.

As usual, I've assumed the motion is confined to one dimension (x). If we allow for a y direction as well, we need a second axis coming out of the page, and the wedges become cones.

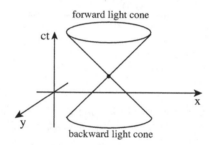

And for all three dimensions (x, y, and z), they are undrawable "hyper-cones": the **forward** and **backward light cones**.

Suppose *you* are the "object" in question. As you live your life, you trace out a world line.[20] At each moment your forward light cone represents the **future** – the region of spacetime you could in principle visit. And because no message you send could travel faster than light, this is the only region you could possibly influence. Similarly, the backward light cone is your **past** – the region of spacetime that can possibly influence you. All the rest is a vast region that is simply inaccessible to you. You can't get there, and you didn't come from there; you can't send a message there, and nobody out there can communicate with you. We call it **elsewhere**.[21]

---

[20] George Gamow entitled his delightful autobiography *My World Line.*

[21] Note that the future, past, and elsewhere are tied to a particular point on your world line. As you move along, your future progressively shrinks, and your past expands; someone else's future is quite different from yours (though they do inevitably overlap).

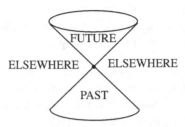

Don't be despondent, though; it's not quite as lonely as it sounds. I may not be able to communicate with you *right now* (you are "elsewhere"), but I certainly *can* talk to the you who will be there in a split second (when your world line enters my forward light cone), and *that* you can answer back to the me who will be here an instant later.

This helps to clarify a very profound problem with causality. Remember that the relativity of simultaneity says that two events that are simultaneous in one inertial frame are not simultaneous in another: A and B happen at the same time for one observer, but B happens before A for another (and for that matter A happens before B for a third). But it is in the very nature of cause and effect that the former must precede the latter. If you could cause something to happen in the past, you could (for example) arrange to poison your infant grandfather. (Think about it . . . . Not a good idea.) The temporal ordering of cause and effect has got to be absolute – otherwise the whole notion of causality is out the window, and with it all of physics. But how is this possible, in light of the relativity of simultaneity?

The answer is that the essential structure of spacetime – the separation into future, past, and elsewhere – *is* absolute, in special relativity. A point in the future is future in *all* inertial systems, and the same goes for past and elsewhere. The relativity of time ordering applies only to events that are elsewhere (ones that are simultaneous in some reference frame). I can cause something to happen in (my) future, and something in (my) past can affect me, but I cannot make anything happen elsewhere (it would require a signal traveling faster than light).

Events happening elsewhere can be earlier, or later, or simultaneous, depending on the observer, but since they are neither causes nor effects of anything I do right now, the ambiguity in their time ordering is innocuous.

Problem 24. By letting the vertical axis be ct, instead of just t, we are, in effect, measuring time in meters. The unit of ct is the distance light travels in one second (a "light-second"). What is a light-second, in meters?

Problem 25. The world line of a certain rocket ship is shown in the following Minkowski diagram.

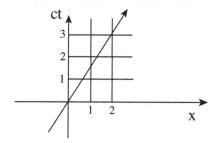

What is the speed of the rocket, (a) as a fraction of the speed of light, and (b) in m/s?

Problem 26. A particle starts at $x = 2\,\text{m}$, at $t = 0\,\text{s}$, and moves in the x direction at a constant speed of $1/5\,c$ for $5 \times 10^{-8}\,\text{s}$. It then stops for $2 \times 10^{-8}\,\text{s}$, turns around, and goes back at constant speed to $x = 0\,\text{m}$. The entire trip took $12 \times 10^{-8}\,\text{s}$.

(a) Sketch the world line for this trip.
(b) On your figure, indicate the particle's "future," at time $t = 6 \times 10^{-8}\,\text{s}$.
(c) For each of the following points in spacetime, say whether it is in the future, the past, or elsewhere, for the particle at $t = 6 \times 10^{-8}\,\text{s}$: (i) $x = 6\,\text{m}, t = 2 \times 10^{-8}\,\text{s}$; (ii) $x = 3\,\text{m}, t = 10 \times 10^{-8}\,\text{s}$; (iii) $x = 2\,\text{m}, t = 7 \times 10^{-8}\,\text{s}$. Plot these points on your graph.

3

Quantum mechanics

Unlike special relativity, quantum mechanics was not the inspired product of one mind. It developed, in fits and starts, over a period of 25 years (1900–1925), and many cooks stirred the broth. Even now, a century later, quantum mechanics raises profound conceptual questions. Every competent physicist can "do" quantum mechanics, and the predictions it makes are in spectacularly good agreement with experimental results; there can be little doubt that quantum mechanics is "right." On the other hand, as Richard Feynman once remarked, "nobody understands quantum mechanics."[1]

I'll tell the story in two parts. First the history (how the essential pieces of the puzzle were assembled), and then the implications (what the finished picture has to say about our world).

3.1 Photons

Like relativity, quantum mechanics began with the study of light. In both cases, this historical association is misleading. Relativity, really, has nothing to do with light – to be sure, it involves that magic speed limit, c, and of course the most familiar thing that travels at speed c is light. So light is a convenient vehicle for introducing the subject. But it is perfectly possible to imagine a universe in which there is no such thing as light, and yet relativity might still be valid.

Quantum mechanics, too, starts with light – in this case the "quantum" of light (the photon). But, as it turns out, this is just one rather specialized example of much more general principles – in fact, a rather *bad* example, because photons are by their nature relativistic (they travel at the speed of

---

[1] Richard Feynman, *The Character of Physical Law*, MIT Press, Cambridge, MA, 1965, p. 129.

69

light), and it would be more natural to begin with the quantum mechanics of slow-moving objects, in analogy with Newtonian mechanics. But that's not how it happened. . . .

3.1.1 Planck's formula

In 1900 Max Planck was struggling with a difficult and intractable problem in the theory of heat radiation – the so-called "blackbody spectrum."[2] He discovered that he could account for the experimental data by assuming that light comes in little "squirts" (which he called **quanta**, but which later acquired the modern name **photons**).

The energy of a photon, he proposed, is proportional to the frequency of the light:

$$E = hf. \tag{3.1}$$

The constant of proportionality h is called **Planck's constant**; to fit the data, Planck found it had to have the value

$$h = 6.626 \times 10^{-34} \, \text{J s}. \tag{3.2}$$

Planck's innocent-looking formula sparked the second great revolution in twentieth-century physics: **quantum mechanics**.

> **Example 1.** For yellow light, $f = 5 \times 10^{14}$ Hz, so the energy of one photon is
>
> $$E = (6.6 \times 10^{-34})(5 \times 10^{14}) = 3 \times 10^{-19} \, \text{J}.$$

If a flashlight puts out, say, 3 watts (=3 J/s), how many photons is it emitting every second?

$$N = \frac{3 \, \text{J/s}}{3 \times 10^{-19} \, \text{J}} = 10^{19} \text{ photons per second.}$$

That's a *lotta* photons! Which is why we never noticed that light comes in these tiny packages. Just as we tend to think of water as a continuous stream, even though we know it actually consists of little particles (molecules), so it is with light: for everyday purposes there are so many photons that it hardly matters. Who's counting?

---

[2] This is the light (or, more generally, electromagnetic waves – it need not be in the visible range) given off by every object, due to the random thermal agitation of its atoms. We'll study blackbody radiation in greater detail in Chapter 5.

Problem 1. Helium–neon lasers have a wavelength of 6.338×10^{-7} m. What is the energy of a single photon in the beam?

Problem 2. How many photons does a 100 watt bulb emit in 1 minute? Assume it is pure yellow light, with a frequency of 5×10^{14} Hz.

3.1.2 The photoelectric effect

Planck did not claim to know *why* light is quantized; he assumed it had something to do with the emission process (some weird thing about the light bulb made it give off light in these little squirts). In 1905 (same year as special relativity) Einstein put forward a far more radical interpretation. He held that Planck's quanta are actual *particles* of light, and Planck's formula had nothing to do with the emission process, but was in the nature of light itself.

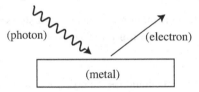

If you shine light on a piece of metal, electrons come popping out. This is called the **photoelectric effect**, and it is the basis for the light meters in cameras, for example. There's nothing surprising about it; metals are *full* of electrons, and I imagine if you hit a piece of metal with a hammer you would knock some of them out. What *is* surprising is that the *energy* of the emerging electrons doesn't depend on the intensity[3] of the light – double the intensity and twice as many electrons come out, but still at the same energy. On the other hand, if you increase the *frequency* of the light, the energy of the emitted electrons *does* increase. Red light yields low-energy electrons; blue light produces high-energy electrons.

Einstein explained all this by assuming that a photon comes in, and delivers its energy (hf) to an electron, which subsequently breaks through the "skin" of the metal (thereby losing energy w, the so-called **work function**), and emerges with energy[4]

$$E = hf - w. \tag{3.3}$$

---

[3] **Intensity** is the energy per unit area, per unit time, carried by the beam. To double the intensity, use two flashlights instead of one, shining on the same spot.

[4] Actually, some electrons bounce around a bit – losing energy – before leaving the metal, so Einstein's formula gives the *maximum* electron energy.

Einstein's formula fits the data perfectly, but for 20 years nobody believed his derivation, because it treats light as a stream of particles. To understand how shocking Einstein's proposition was, you have to recall the old debate about whether light is a wave or a particle phenomenon. Young's double-slit experiment had settled the issue definitively in favor of the wave model, and Maxwell's electrodynamics provided a compelling and comprehensive theoretical explanation. Then here comes Einstein (just 25 years old) saying that, after all, light is actually a stream of particles! How preposterous.

When Planck and others nominated Einstein for membership in the Prussian Academy, in 1913, they wrote,[5]

> In sum, one can say that there is hardly one among the great problems in which modern physics is so rich to which Einstein has not made a remarkable contribution. That he may sometimes have missed the target in his speculations, as, for example, in his hypothesis of light-quanta, cannot really be held too much against him, for it is not possible to introduce really new ideas even in the most exact sciences without sometimes taking a risk.

It was not until 1923 that the photon was taken seriously as a "particle" of light. What turned the tide was the **Compton effect**.

Problem 3. Light of wavelength 4.5×10^{-7} m strikes a piece of metal, knocking out electrons. The work function for this metal is 2×10^{-19} J. (a) What is the energy of the incoming photon? (b) What is the (maximum) energy of the emitted electrons?

3.1.3 The Compton effect

When light reflects off a mirror it changes its *direction*, of course, but not its *color* (frequency). But when light scatters off an individual electron, it *does* change color. In 1923, Arthur Compton was studying this effect in the laboratory, and found that he could explain his observations by treating the process as an ordinary (relativistic) collision between two *particles*, a (massless) photon and an electron, with the photon carrying energy given by Planck's formula (Eq. 3.1) and momentum given by Eq. (2.15):[6]

$$E = hf, \qquad p = \frac{E}{c} = \frac{hf}{c} = \frac{h}{\lambda}. \tag{3.4}$$

---

[5] Quoted in Abraham Pais, *Subtle is the Lord...*, Oxford University Press, New York, 1982, p. 382.

[6] To avoid notational clutter, I'll drop the subscript r on relativistic energy and momentum from now on.

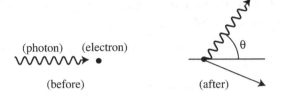

(photon)   (electron)

(before)   (after)

Using the conservation of energy and momentum, he calculated the energy (hence the frequency) of the outgoing photon, as a function of the scattering angle θ, and his formula fitted the data perfectly. The Compton effect forced physicists to accept Einstein's interpretation: at the microscopic level, light really *does* behave like particles.

Problem 4. Suppose the photon in Compton scattering bounces directly back ($\theta = 180°$). Show that

$$\lambda' = \lambda + \frac{2h}{mc},$$

where λ is the wavelength of the incoming photon, λ' is the wavelength of the outgoing photon, and m is the (rest) mass of the electron. [This is tricky, so I'll help you get started: in terms of their wavelengths, find the energy and momentum of the incoming and outgoing photons. What are the (relativistic) energy and momentum of the electron, before the collision? Call the energy and momentum of the outgoing electron E and p. Write down the equations for conservation of energy and conservation of momentum (note that the final photon is going to the *left* – its momentum is negative). Solve these equations for E and p, and plug the result into Eq. (2.14).]

3.1.4 de Broglie's hypothesis

The following year (1924), a young Frenchman, Louis de Broglie, made a reckless but inspired conjecture: if light behaves both as a wave and as a particle, perhaps ordinary particles (electrons, for instance) have wavelike properties. Specifically, he proposed that *particles have an associated wavelength*, given by the formula

$$\lambda = \frac{h}{p}, \tag{3.5}$$

where p is the particle's momentum (this just inverts the formula Compton used for the momentum of a photon in terms of its wavelength, Eq. (3.4)). We call λ the **de Broglie wavelength** of the particle.

Example 2. What is the de Broglie wavelength of a marble (10 grams) going at 2 m/s?

Solution: The momentum of the marble is $p = (0.01)(2) = 0.02 \, \text{kg m/s}$, so

$$\lambda = \frac{6.63 \times 10^{-34}}{0.02} = 3.32 \times 10^{-32} \, \text{m}.$$

That's *awfully* small – an atom is about 10^{-10} m in diameter, and the nucleus is about 10^{-15} m across. To get a larger wavelength we would need the smallest possible mass (the electron, say, at 9.11×10^{-31} kg), going as slowly as possible. At room temperature the thermal velocity of an electron is about 10^5 m/s, so

$$\lambda = \frac{6.63 \times 10^{-34}}{9.11 \times 10^{-26}} = 7 \times 10^{-9} \, \text{m}.$$

That's still pretty small – comparable to the size of an atom. You could cool the electrons down, so they're not moving as fast, but the moral of the story is that you're going to need something very light and very slow, interacting with something extremely small (such as an atom), before the wave nature of the particle will become apparent.

Einstein championed **de Broglie's hypothesis** (once again in the face of widespread skepticism), and in 1925 Davisson and Germer inadvertently verified it in the laboratory. Remember, the classic test of a wave phenomenon is interference – in the case of light, Young's double-slit experiment. You pass a beam through two narrow slits, and it splits into a number of separate beams. If you do it with multiple slits, it is called **diffraction**. Davisson and Germer found that when you fire a beam of electrons at a crystal, the regular array of atoms acts like a bunch of slits, and the beam splits into several, just as it would for light.[7] This **electron diffraction** confirmed de Broglie's hypothesis that particles behave in some respects as waves.

Evidently light, which we had thought was a wave, has particle-like attributes, and electrons, which we have always regarded as particles, have a wave-like aspect. This **wave–particle duality** was utterly mysterious at the time, and in some respects it remains so to this day. It was the inspiration for Heisenberg, Schrödinger, Born, and others, who built the modern theory of quantum mechanics in the years 1925–1927. But before we come to that, I

---

[7] Actually, for diffraction off a crystal you need X-rays, which have much shorter wavelengths than visible light – comparable, in fact, to the de Broglie wavelength of Davisson and Germer's electrons.

must step back and pick up another strand in the story: Bohr's theory of the hydrogen atom.

Problem 5. Find the de Broglie wavelength of a baseball (mass 2 kg) going at 33 m/s.

Problem 6. If an electron has the same wavelength as yellow light, how fast is it going?

3.2 The Bohr model

By about 1910, experiments in Ernest Rutherford's laboratory indicated that the atom consists of light-weight, negatively charged electrons orbiting a tiny but much heavier, positively charged nucleus, rather like planets going around the Sun. But the laws of classical electrodynamics dictate that a particle going in a circle should radiate (give off light), thus losing energy, and hence that the electron should spiral quickly (in fact, in a tiny fraction of a second) into the nucleus. If that were true, atoms wouldn't last long.

What is more, the light that is emitted and absorbed by an atom does not come in a broad range of wavelengths, but only in specific colors. Each chemical element (hydrogen, carbon, oxygen, etc.) has its own **spectrum** of frequencies – the unique electromagnetic fingerprint of that particular atom. For example, the visible portion of the spectrum of hydrogen includes five distinct colors.[8]

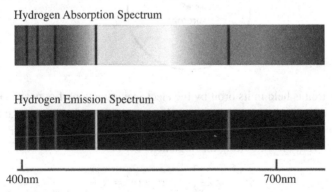

Hydrogen Absorption Spectrum

Hydrogen Emission Spectrum

400nm   700nm

---

[8] In the figures, the light has been spread out like a rainbow, by a prism, to reveal the colors it contains. The **absorption spectrum** is obtained when you pass white light (all colors) through a container of hydrogen; the frequencies at which hydrogen absorbs light show up as dark lines. To get the **emission spectrum** you heat up the gas, and examine the light it emits. Emission and absorption spectra are complementary – dark lines in absorption correspond to bright lines in emission.

In 1913, Niels Bohr – using a serendipitous amalgam of inapplicable classical physics and premature quantum mechanics – proposed a model of the hydrogen atom (the simplest of all atoms, with a single electron in orbit around a nucleus consisting of a single proton). In Bohr's vision the electron is not allowed to be in just *any* old orbit, but only in those orbits such that the de Broglie wave fits perfectly around the circumference (either *one* full wavelength, or *two*, or *three*, etc., but nothing in between). This is reminiscent of the condition for standing waves on a string – they reinforce constructively only when the wavelength is "just right" to join up in phase after executing one orbit.[9]

Bohr decreed (he didn't pretend to offer any justification for this) that electrons in these **allowed orbits** do not radiate, but when they make a transition (a **quantum jump**) from one allowed orbit to another, they emit or absorb a photon, which carries the difference in energy between the two states. His theory accounted perfectly for the observed spectrum of hydrogen – something that was utterly mysterious at the time. Bohr's result remains (to my mind) the greatest triumph of quantum mechanics.

3.2.1 Allowed energies

Here, then, is Rutherford's picture of the hydrogen atom.

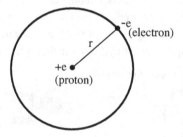

The electron is held in its orbit by the electrical attraction of opposite charges (Eq. (1.17)): ke^2/r^2, where e is the charge of the proton (the charge of the electron is $-e$). Newton's second law says $F = ma$, and for circular motion $a = v^2/r$ (Eq. (1.9)). Putting this together,

$$m\frac{v^2}{r} = k\frac{e^2}{r^2}, \quad \text{or} \quad mv^2 = k\frac{e^2}{r}. \tag{3.6}$$

---

[9] I'm doing some violence to history, you'll notice, since de Broglie didn't introduce his hypothesis until 11 years later. Bohr's argument was more convoluted than this, but telling the story in terms of de Broglie waves makes it a little more plausible.

(This is just like planetary motion, except that the force is electrical instead of gravitational.)

The kinetic energy of the electron is[10]

$$T = \frac{1}{2}mv^2 = \frac{1}{2}\frac{ke^2}{r}, \tag{3.7}$$

and its potential energy (Eq. (1.30)) is

$$V = -k\frac{e^2}{r} \tag{3.8}$$

(negative, since opposite charges attract – you would have to *do* work to pull them apart). The total energy is therefore

$$E = T + V = -\frac{1}{2}\frac{ke^2}{r}. \tag{3.9}$$

That's all *classical* physics. Now de Broglie says

$$p = mv = \frac{h}{\lambda}, \quad \text{so} \quad \lambda = \frac{h}{mv}, \tag{3.10}$$

and Bohr stipulates that a whole number (n) of wavelengths fit around the circumference

$$n\lambda = 2\pi r \quad (n = 1, 2, 3, \ldots). \tag{3.11}$$

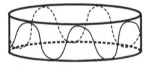

Thus

$$n\frac{h}{mv} = 2\pi r, \quad \text{or} \quad v = \frac{nh}{2\pi r m}.$$

But from Eq. (3.6) we have $v^2 = ke^2/mr$, so

$$\frac{ke^2}{mr} = \frac{n^2h^2}{4\pi^2r^2m^2},$$

or, solving for r (and affixing a subscript to indicate the value of n)

$$r_n = \left(\frac{h^2}{4\pi^2kme^2}\right)n^2 = r_1 n^2 \quad (n = 1, 2, 3, \ldots). \tag{3.12}$$

---

[10] As we shall see, the speed of an electron in hydrogen is much less than c, so it is appropriate to use the nonrelativistic expressions for energy and momentum.

If this is true, the electron's orbit cannot have just any old radius, but only certain "allowed" values:

$$r_1 = \frac{h^2}{4\pi^2 kme^2} = 5.292 \times 10^{-11}\,\mathrm{m}, \tag{3.13}$$

which is called the **Bohr radius**, or

$$r_2 = 4r_1 = 2.117 \times 10^{-10}\,\mathrm{m},$$

or

$$r_3 = 9r_1 = 4.763 \times 10^{-10}\,\mathrm{m},$$

and so on. And the same goes for energy (Eq. 3.9): the electron cannot have just any old energy – only the **allowed energies** given by Bohr's famous formula

$$E_n = -\left(\frac{2\pi^2 k^2 me^4}{h^2}\right)\frac{1}{n^2} = -\frac{E_1}{n^2} \quad (n = 1, 2, 3, \ldots). \tag{3.14}$$

The lowest possible energy occurs when $n = 1$ (the **ground state**):[11]

$$E_1 = -\frac{2\pi^2 k^2 me^4}{h^2} = -13.61\,\mathrm{eV}. \tag{3.15}$$

If the atom is in its ground state, it would take 13.6 eV to strip the electron off ("ionize" the atom). In the **first excited state** ($n = 2$), with energy

$$E_2 = \frac{E_1}{4} = -3.40\,\mathrm{eV}$$

it would take 3.40 eV to ionize the atom, and so on.

If the electron makes a quantum jump[12] from one allowed orbit (n_i) to another (n_f) of lower energy ($n_f < n_i$), the energy lost is carried off by an emitted photon:

$$E_{\text{photon}} = E_{\text{initial}} - E_{\text{final}} = -13.6\left(\frac{1}{n_i^2} - \frac{1}{n_f^2}\right)\,\mathrm{eV}. \tag{3.16}$$

If the final energy is *higher* than the initial energy, then the atom must *absorb* a photon to make up the difference. The frequency of the photon is given, in either case, by Planck's formula,

$$E_{\text{photon}} = hf. \tag{3.17}$$

In the figure below I have indicated some of the possible transitions. Transitions to the lowest-energy state ($n_f = 1$) yield photons in the ultraviolet – this

---

[11] An electron volt, eV, is 1.602×10^{-19} J, a convenient unit of energy in atomic physics.

[12] Why, how, and when the quantum jumps occur are questions the Bohr model does not address.

is called the **Lyman series**; transitions to the state $n_f = 2$ (the **Balmer series**) are in the visible range; transitions to $n_f = 3$ (the **Paschen series**) are infrared, and so on.

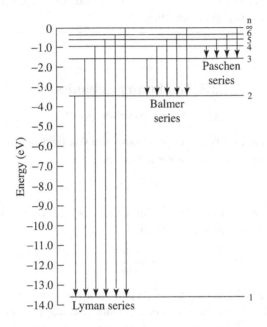

Problem 7. Put in the known values of h, k, m, and e to determine (a) the Bohr radius (Eq. (3.13)) and (b) the ground state energy of hydrogen (Eq. (3.15)) – first in joules, then convert to electron volts.

Problem 8. A hydrogen atom undergoes a transition from the state $n = 4$ to the state $n = 1$. (a) What was the initial energy of the atom? (b) What is the final energy of the atom? (c) What is the energy of the emitted photon? (d) What is the frequency of the emitted photon? (e) Is this photon in the visible region? If not, what sort of radiation is it?

Problem 9. An electron in a hydrogen atom makes a transition from the state $n = 4$ to the state $n = 2$. (a) What is the energy of the emitted photon? (b) What is its frequency? (c) What is its wavelength?

Problem 10. What is the kinetic energy (Eq. (3.7)) of the electron in the ground state of hydrogen? What is its velocity? What percent of the speed of light is this? Would you agree that it is reasonable to use the classical formulas for energy and momentum?[13]

3.3 Quantum mechanics

3.3.1 Wave–particle duality

As we have seen, Planck and Einstein proposed that light (a wave) behaves in some ways as a particle (the photon), with energy

$$E = hf.$$

Later de Broglie suggested that particles (such as electrons) behave in some respects like waves, with a wavelength

$$\lambda = \frac{h}{p}$$

(where $p = mv$ is the momentum of the particle). This "wave–particle duality," which Bohr elevated to the status of a cosmic principle (**complementarity**), makes it sound as though photons and electrons are like unpredictable adolescents, who sometimes act as adults, and at other times, for no apparent reason, as children. It is true that the simple dichotomy between waves and particles is subtle, in quantum mechanics, but roughly speaking, light is a wave when (as in a flashlight beam) enormous numbers of photons are involved, and its particle nature comes into play only when (as in the photoelectric effect or Compton scattering) individual photons interact with individual electrons. And the wave nature of a particle is relevant only when λ is comparable in size to the objects with which the particle interacts.

3.3.2 The wave function

de Broglie's hypothesis raises an obvious question: what *is* this wave? How does it propagate, and what is its connection to the particle? De Broglie didn't have a clue. For want of a better name it was called, simply, the **wave function**, and assigned the Greek letter psi: Ψ.

---

[13] Relativity does make a correction to the Bohr theory, but it is very small.

In 1925, Schrödinger proposed an equation for calculating Ψ:

$$i\hbar\frac{\partial\Psi}{\partial t} = -\frac{\hbar^2}{2m}\frac{\partial^2\Psi}{\partial x^2} + V\Psi. \tag{3.18}$$

I'm not going to explain all the terms in **Schrödinger's equation** – you'll notice that the mass of the particle (m) comes into it, and also the potential energy (V). Schrödinger's equation is the fundamental law of quantum mechanics. It plays a role logically analogous to Newton's second law ($F = ma$) in classical mechanics. Newton's law tells you how the particle moves, under the influence of a given force; in the same sense, the Schrödinger equation tells you how its wave function evolves.[14] The goal of classical mechanics is to predict the *position* of the particle at any future time; the goal of quantum mechanics is to determine the *wave function* of the particle at any future time.

Very well... but what does this "wave function" tell us about the particle? Schrödinger had no idea, and when (in 1926) Max Born supplied the answer, Schrödinger didn't believe it.[15]

3.3.3 Born's statistical interpretation

The wave function, as its name suggests, is spread out in space:

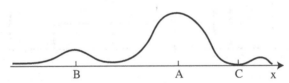

How can such a thing describe a particle, which by its nature is localized at a point? According to Born, Ψ (or rather, its square, Ψ^2) tells you the *probability* of finding the particle at a particular point, were you to make a measurement.[16] In the case of the wave function in the figure, you would be quite likely to find the particle in the vicinity of point A, rather unlikely to find it near B, and certain *not* to find it at C. (As time goes by, the wave function changes shape, in accordance with Schrödinger's equation, so a moment later you might be *un*likely to find the particle at A and likely to find it at C.) More precisely, the probability of finding the particle between point a and point b is the area under the graph of Ψ^2, between a and b.

---

[14] In effect, the potential energy is another way of specifying the force. In classical mechanics it is simpler to work with F, but in quantum mechanics it is easier to work with V.

[15] It's an interesting historical fact that three of the principal founders of quantum mechanics, Einstein, de Broglie, and Schrödinger, never fully accepted the theory in its final form.

[16] Actually, Ψ is a "complex" number, with real and imaginary parts; it is the *absolute* square ($|\Psi|^2$) we need here. But this is a technical detail that is irrelevant to our purpose, and I shall think of Ψ as an ordinary (real) number.

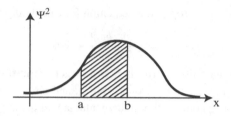

Example 3. Suppose the wave function of a particle is given in Figure (a); Ψ^2 is shown in Figure (b). In that case the probability of finding the particle between 0 and 1 is 1/4, and between 1 and 2 it is 1/4; it is certainly *not* to be found between 2 and 3.5, and the probability of finding it between 3.5 and 4 is 1/2. Note that the total is 1(certainty) – it's gotta be *somewhere*.

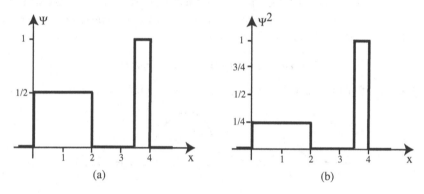

(a)   (b)

Problem 11. The wave function for a particle is shown in the graph below. (a) What is the probability that the particle would be found between $x = 1$ and $x = 2$? (b) What is the total probability of finding the particle somewhere between $x = 0$ and $x = 5$?

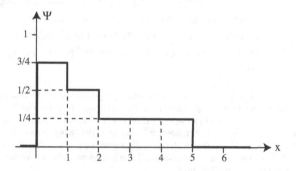

3.3.4 Indeterminacy

Born's statistical interpretation introduces a kind of **indeterminacy** into quantum mechanics, for even if we know everything the theory has to tell us about the particle (to wit: its wave function, Ψ), still we cannot predict with certainty the outcome of a simple experiment to measure its position. All quantum mechanics has to offer is *statistical* information about the *possible* outcomes. This indeterminacy has troubled physicists and philosophers alike.[17] It is natural to wonder whether it is a fact of nature, or a defect in the theory. I'll return to this question in Section 3.4.

3.3.5 Uncertainty

I suggested a moment ago that a wave by its nature is spread out in space, but just *how much* it spreads out depends on the particular case. Here's a wave that is spread out a *lot*:

If somebody asked you "Where *is* that wave?" you would probably think he was a bit nutty, because that wave isn't really *anywhere* – it's kind of smeared out. On the other hand, for this wave:

it is *not* so nutty to ask where it is – it's pretty well localized around the point A.

But what if I had asked instead, "What is the *wavelength* of that wave?" We could give a pretty definite answer for the first wave, but the question is nutty for the second, since it doesn't complete even one full oscillation. The wavelength, remember, is the distance between adjacent peaks, and in this case there is only one peak.

---

[17] Einstein famously objected that "God does not play dice." This is actually a common paraphrase of what he wrote in a letter dated Dec. 4, 1926, to Born. (*The Born–Einstein Letters*, Walker, New York, 1971.) When he repeated the assertion at the fifth Solvay Conference, in 1927, Bohr supposedly retorted, "Einstein, stop telling God what to do." Some philosophers seem to think that the statistical interpretation rescues free will from the tyranny of classical determinism. Actually, at best it substitutes the even more onerous tyranny of chance. Free will has nothing to do with it.

It is an inescapable trade-off: If you want to know the position of a wave, it's got to be sharply localized, and in that case it cannot have a well-defined wavelength; if you want to know the wavelength, then it's got to be spread out, and in that case it doesn't have a well-defined position. There is nothing particularly deep about this, and it has nothing to do with quantum mechanics – it's just a (pretty obvious) fact about waves.

But in quantum mechanics the wavelength is related (by de Broglie's formula) to the *momentum* of the particle ($\lambda = h/p$), and it follows that if you know the *position* of a particle reasonably precisely, you cannot know much about its *momentum* – and conversely, if the *momentum* is known to precision, then the wave function must extend over a substantial distance, and in that case its *position* is ill-defined (a measurement could return a wide range of answers).

This realization takes quantitative form in Heisenberg's **uncertainty principle**:[18]

$$(\Delta x)(\Delta p) > h \qquad\qquad (3.19)$$

(where Δ, in this case, is to be read as "the uncertainty in . . . "). In classical physics you could, in principle, measure both the position and the momentum of a particle to arbitrary precision. In quantum mechanics this is limited by the uncertainty principle: the smaller the uncertainty in position (Δx), the greater is the uncertainty in momentum (Δp), and vice versa.[19]

"But that's absurd!" (I hear you say). "I can measure the position of a particle, and however spread out the wave function may have been, I do in fact find it at one point. And if I then decide to measure its momentum, I will again get a specific result." This is quite true, but the inference you are hoping to draw ("I know both the position and the momentum of the particle to arbitrary precision."), is false. For if you now go back and *re*measure the position, you will get a completely different answer! In the act of measuring the momentum, you inevitably (and randomly) alter the position, and vice versa.

"Nonsense! I will be exquisitely careful not to change anything." Unfortunately, in order to measure the position of a particle, you have to *do* something to it – shine a light on it, for example. That means hitting it with a photon, and the resulting collision is going to alter its momentum. That's how the uncertainty principle is enforced by the process of measurement.[20] In the macroscopic

---

[18] The symbol > means "greater than."

[19] Notice what "uncertainty" means, in this context. It refers to the *range* of values you might get, if you measured the quantity in question.

[20] In the famous **Bohr–Einstein debates**, Einstein tried to devise experiments that would violate the uncertainty principle, but Bohr was always able to demonstrate – sometimes brilliantly – that the details of the measurement would in fact sustain Heisenberg's formula.

world you can (in principle) make a measurement without significantly affecting the object you're measuring, but in the microscopic domain this is no longer possible; the uncertainty principle puts an absolute limit on the precision with which you can measure (and hence *know*) both the position and the momentum of a particle.[21]

Problem 12. An electron (mass 9×10^{-31} kg) is localized to within 1 mm. What is the (minimum) uncertainty in its velocity?

Problem 13. A baseball, with a mass of 0.5 kg, is known to be inside a shoebox 30 cm long. (a) What is the (minimum) uncertainty in its momentum? (b) What is the (minimum) uncertainty in its velocity? (c) At this speed, how long would it take to get from one end of the box to the other? (d) In view of this result, how relevant is the uncertainty principle for macroscopic objects?

Problem 14. In Bohr's model, the electron in the ground state of hydrogen is confined to an orbit of radius r_1 (Eq. (3.13)). At one moment $x = +r_1$, and half an orbit later $x = -r_1$, so let's say $\Delta x = 2r_1$. Find the momentum of the electron (Eq. (3.10)), p_1. In the same spirit, let's say $\Delta p = 2p_1$. Are these values consistent with the uncertainty principle?[22]

3.3.6 Tunneling

An interesting implication of the Schrödinger equation is the phenomenon of **tunneling**. Picture a frictionless, motorless roller coaster approaching a hill:

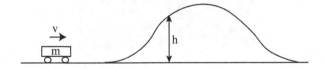

---

[21] Note that there is no limit on the precision with which you can measure *just* the position, or *just* the momentum. It's when you try to measure *both* of them that you get into trouble with the uncertainty principle. By the way, there are other pairs of **incompatible observables** in quantum mechanics, linked by other "uncertainty principles."

[22] Don't take this result too seriously – these are just *estimates* of Δx and Δp, and Bohr theory is not rigorous quantum mechanics.

As it rides up the incline, it slows down; its kinetic energy decreases, while its potential increases – the total remaining constant (that's conservation of energy). The coaster will come to a stop when its potential energy (mgh) equals its original kinetic energy ($\frac{1}{2}mv^2$):

$$mgh = \frac{1}{2}mv^2 \quad \Rightarrow \quad h = \frac{v^2}{2g}. \tag{3.20}$$

Classically, if the hill is higher than this, the roller coaster cannot make it over the top; it slides back down in the direction from which it came.

But quantum mechanically the wave function does not go to zero at the classical turning point. It penetrates into the hill, and a small fraction leaks out the other side. This is simply a fact about solutions to the Schrödinger equation; physically, it means that the particle has a finite probability of "tunneling" through the barrier. Quantum tunneling accounts for certain kinds of radioactivity, and it has been exploited in electronics and microscopy.

A recent TV program claimed that quantum tunneling means you can walk through walls. For some reason it never works when I try it – though this fellow seems to have made it part way:[23]

Like other quantum phenomena, tunneling does not really apply to macroscopic systems like roller coasters and people – the probabilities are so utterly minute that the whole idea is a joke. We invoke these macroscopic analogies only to illustrate the *principle* involved.

---

[23] This sculpture in Paris was inspired by Marcel Aymé's 1943 short story *Le Passe-Muraille* (*The man who could walk through walls*).

Problem 15. A roller coaster going 10 m/s approaches a hill 6 m high.

(a) Classically, does it make it over the top? If not, at what height does it stop and reverse direction?

(b) Quantum mechanically, could it make it to the other side?

3.4 What's so funny about quantum mechanics?

Schrödinger's equation and Born's statistical interpretation of the wave function[24] finally put quantum mechanics on a sound theoretical foundation. Schrödinger himself solved his equation using the Coulomb potential (Eq. (3.8)), and recovered Bohr's equation for the allowed energies of the hydrogen atom (Eq. (3.14)). But unlike Bohr's argument, which relied on inapplicable classical formulas and questionable ad hoc assumptions, Schrödinger's derivation was based on rigorous mathematical deduction. The electron doesn't follow circular classical orbits (as Bohr supposed); rather, Ψ^2 describes an **electron cloud**, with regions where the electron is more and less likely to be found:

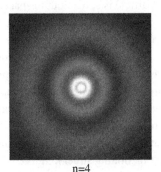

n=2   n=4

The Schrödinger equation is not consistent with special relativity, however, and in 1927, P. A. M. Dirac introduced a relativistic version. Dirac's equation had the astonishing implication that the electron (or any other particle, for that matter) has an **antiparticle** twin, with the same mass, but opposite charge. No such particle was known at the time, and Dirac struggled at first to explain it

---

[24] In its original form, the statistical interpretation tells you how to extract (probabilistic) information about the *position* of the particle from Ψ, but it was soon generalized to other observables, such as momentum, energy, and so on.

away.[25] But in 1932 Carl Anderson (who was only dimly aware of Dirac's prediction) discovered the **positron** in an experiment on cosmic rays. It was the triumphant offspring of the marriage of quantum mechanics and special relativity.

Ironically, the photon – which initiated the whole quantum saga back in 1900 – did not have a completely satisfactory theory until the 1940s, when Richard Feynman, Julian Schwinger, and Sin-itiro Tomonaga put the final touches on **quantum electrodynamics**.

Quantum mechanics accounts for the structure of atoms and molecules, and it explains the **Periodic Table** of the elements; it is the reason why solids are solid, and stars do not (ordinarily) collapse. It led to the invention of the transistor, the electron microscope, and the laser. It provides the mechanism for superconductivity and ferromagnets. In truth, there is hardly a phenomenon in nature that is not touched in some way by quantum mechanics. It is certainly *correct*.

And yet... quantum mechanics raises profound and perplexing conceptual questions. That, indeed, is what makes the subject so fascinating. There are several distinct issues involved, but the underlying problem is indeterminacy: quantum mechanics cannot predict with certainty the outcome of a measurement, and as a result, it does not provide a coherent and compelling picture of physical reality. This is true in the first instance for *position*. If the wave function of a particle is spread out in space, then where exactly *is* the particle? Is it in many places at once (whatever that means)? Is it nowhere at all? You might be tempted to conclude that there just isn't any such thing as a particle – it's all waves. But then how do you account for the fact that a measurement always returns a particular location? A photon may be "smeared out" in advance of a measurement, but when you hold up a piece of film, it leaves a spot at just one point; an electron may have a broad wave function, but when it hits a TV screen only one pixel lights up.

And it's not just *position*; the same thing applies to *all* observables. A properly functioning street light is either red, yellow, or green, at any particular moment. But in quantum mechanics its wave function could be a combination of all three;[26] and yet, when you make a measurement you only get one of them. In any given context quantum mechanics can tell you the *probability* of a particular outcome... but somehow we expect more than that from a fully satisfactory physical theory. Is quantum indeterminacy simply a matter of *ignorance*, or is it telling us something deep about the world?

---

[25] His original proposal – that the antielectron is the proton – fails for the obvious reason that it is 2000 times too heavy.

[26] Of course, this doesn't apply to macroscopic macroscopic microscopic objects like street lights – they're either red, yellow, or green, *period*, no probabilities involved. But it *does* happen in the microscopic realm, and one of the enduring puzzles of quantum mechanics is how and why the quantum world of probabilities becomes the classical world of determinism, as a system gets larger and more complex.

3.4.1 Three schools of thought

Suppose I have a particle represented (at some instant t) by the following wave function:

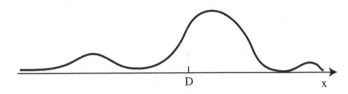

I measure the position of the particle, and I find it to be at the point D. *Question:* Where was the particle immediately *before* my measurement?[27] There are three plausible responses, and they serve to characterize the main schools of thought regarding quantum indeterminacy:

(1) The **realist** response: *The particle was at D.* This certainly seems sensible, and it is the answer advocated by Einstein. However, if this is true then quantum mechanics is an *incomplete* theory, since the particle *really was* at D, but quantum mechanics didn't know it. Evidently Ψ is not the whole story – some additional information (known in the business as a **hidden variable**) would be required, to complete the description.[28]

(2) The **orthodox** response: *The particle wasn't really anywhere* – the act of measurement forced it to "take a stand" (though how and why it decided on the point D we dare not ask). This view (the so-called **Copenhagen interpretation**) is associated with Bohr; among physicists it has always been the most widely accepted view. But if this is correct, there is something very bizarre about the measurement process – something that nearly a century of debate has done precious little to illuminate.

It gets worse. What if I made a *second* measurement, immediately after the first: Would I get D again, or would the second act of measurement generate a completely new result? On this question practically everyone is in agreement: a repeated measurement must return the *same* value. (Indeed, it would be pretty meaningless to say that the particle was found to be at D in the first instance, if this could not be confirmed by immediate repetition of the measurement.) To the realist there is nothing surprising about this – the particle *really was* at D before the first measurement, you found it

---

[27] It's got to be *immediately* before, so the particle hasn't had a chance to move around, or the wave function to change its shape.

[28] Note that **realism** is a technical term, here: it implies that the particle had well-defined properties prior to measurement. The alternative is not *mysticism*, as some authors like to suggest, but "orthodoxy," as explained in the next paragraph.

to be at D, and when you checked it was *still* at D. But how does the orthodox interpretation account for the fact that the second measurement is bound
to return the same value? It must be that the first measurement precipitates
a **collapse of the wave function**, so that it is now sharply peaked about the
point D:

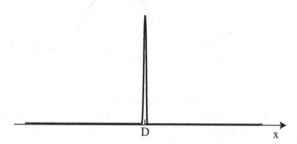

In the orthodox interpretation, then, not only does the act of measurement
force the particle to "take a stand," but it instantaneously collapses the wave
function, thereby guaranteeing that a repetition of the measurement will
yield the same result.

(3) The **agnostic** response: *Refuse to answer.* This is not as silly as it sounds.
After all, what sense can there be in making assertions about the position
of the particle before a measurement is made, when the only way to check
whether you were right is precisely to *make* a measurement, in which case
what you got was not "before the measurement"? This was the position
advocated by Wolfgang Pauli, who argued that the question is metaphysical,
in the pejorative sense of the word – like asking how many angels can dance
on the head of a pin. Physics should confine its attention to things that can
be measured, and not make assertions about matters that cannot, by their
nature, be verified in the laboratory. Until about 1965 this was the fall-back
response for most physicists: if you objected to the orthodox answer they
would retreat to the agnostic response, and terminate the conversation.

Problem 16. For each of the following terms, give a brief (2–3 sentence)
explanation, in your own words, of what it means in quantum mechanics.

(1) Indeterminacy.
(2) Uncertainty.
(3) Realism.
(4) Copenhagen interpretation.
(5) Collapse (of the wave function).

3.4.2 The EPR paradox

In 1935 Einstein, Podolsky, and Rosen published the famous **EPR paradox**, which was designed to prove that the realist position is the only tenable one. I'll describe a simplified version, introduced by David Bohm. Consider the decay of a neutral pi meson into an electron and a positron:

$$\pi^0 \rightarrow e^- + e^+.$$

Assuming the pion was at rest, the electron and positron fly off in opposite directions:

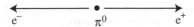

Now, the pion has spin zero, while the electron and positron have spin one-half, so conservation of angular momentum dictates that if the electron comes out with spin up, the positron must have spin down, and vice versa.

"*Stop!* What's all this about pi mesons, and spin, and angular momentum ... ?" OK. Forget about the pion and the electron and the positron. Imagine instead that I have two frisbees, one in each hand, and I toss them to my friends Mary and Tom. By the nature of the way I launch them, one is spinning clockwise (viewed from above) and the other counterclockwise:

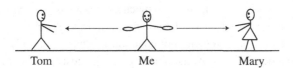

But now suppose I am shrouded in smoke, and I turn around a random number of times before throwing the frisbees. It is still the case that one of my friends gets a clockwise frisbee, and the other gets a counterclockwise frisbee, but they won't know until the frisbees emerge from the smoke who gets which.

Suppose Mary catches her frisbee, and notes that it is spinning clockwise. *Immediately* she knows that Tom's frisbee is spinning counterclockise. To the realist there is nothing peculiar about this – Mary's frisbee *really was* clockwise (and Tom's counterclockwise) from the moment of launch. But in the orthodox interpretation *neither one* was clockwise *or* counterclockwise until the act of measurement forced it to "take a stand." (Of course, this doesn't actually apply to macroscopic objects like frisbees, which contain enormous numbers of atoms – that's why Bohm spoke about pions, electrons, and positrons. But the *logic* is the same.) So when Mary examined her frisbee, a "message" must have gone out to Tom's frisbee, saying, "For God's sake be counterclockwise this time – otherwise we're about to break the rules." Schrödinger called this situation, in which a measurement on one particle determines the properties of some other particle, an **entangled state**.

You're probably picturing Mary and Tom being, say, 10 feet apart, but in principle they could be 10 *light years* apart, for all I care. Nevertheless, the "message" has got to get to Tom *instantaneously* – much faster than the speed of light. Otherwise Tom could measure his frisbee before the message had arrived, and in that case both of them might come out clockwise. Whatever you think of this in the abstract, the experiments have been done, and the results are unequivocal: the spins are always opposite, even if no signal traveling at the speed of light (or less) could possibly have got there in time. The "vehicle" that carries the message from Mary to Tom is precisely the collapse of the wave function. Evidently this collapse has to be instantaneous, even over large distances.

Einstein called this "spooky action-at-a-distance" (the sophisticated term is **nonlocality**),[29] and he regarded it as so preposterous that the orthodox interpretation could not be right: the frisbees must have had their particular spins from the moment of launch. They were not acquired in the act of measurement.

3.4.3 Bell's theorem

Well, if Einstein is right, then quantum mechanics is an *incomplete* description of physical reality: particles have definite positions (and electrons have definite spins) prior to any measurement, but quantum mechanics doesn't know what they are. The indeterminacy of quantum mechanics is just a matter of ignorance. In that case the urgent task, obviously, is to *complete* the theory – to find the "extra information" (the hidden variable) that, together with Ψ, would enable us to predict with certainty the outcome of any experiment. You would think that the EPR paradox would set off a race to find the correct hidden variable theory. But that's not what happened.

In 1932, John von Neumann had published an abstruse proof that no deterministic theory could be compatible with quantum mechanics. His argument contained an unjustified assumption, but that was not discovered until many years later; in the mean time his "proof" certainly dampened enthusiasm for the project. And in 1935 Bohr wrote an unintelligible rebuttal to the EPR paradox. I doubt that many people actually read it, but they were relieved to know that the great man had taken care of the problem. In 1952, David Bohm produced

---

[29] A nonlocal influence is one that propagates faster than the speed of light.

a hidden variable theory that *is* compatible with quantum mechanics (so much for von Neumann's proof); it is cumbersome and implausible, and it includes spooky action-at-a-distance (the very thing Einstein was trying to avoid), but it is fully deterministic, and to this day it has its adherents. But in the 30 years after the EPR paradox nobody managed to come up with a *local* hidden variable theory.

Finally, in 1964 John Bell proved that *any local realist theory is incompatible* with quantum mechanics.[30] This was way beyond anything Einstein had suggested; he took it for granted that quantum mechanics is correct, as far as it goes – he just thought it is not the whole story. Bell's theorem effectively eliminated the agnostic position; it makes an *observable difference* whether the particle had a particular spin (or location) prior to measurement, or not – even if we do not know what that spin (or location) *is*. Pauli was simply mistaken, in asserting that it is a metaphysical question. Most important, Bell's theorem turned it into an *experimental* question whether quantum mechanics, or local realism, is the correct theory.

Over the next 20 years the experiments were performed, culminating in the definitive work of Alain Aspect, in 1982. Suffice it to say that they decisively confirmed the quantum predictions, and ruled out any possibility of local determinism. Indeterminacy is *not* a result of our ignorance or incompetence – it appears to be a fact of life.[31] And "spooky action-at-a-distance," which Einstein considered so preposterous, is inescapable.

3.4.4 Nonlocality

But doesn't special relativity tell us that nothing can travel faster than light? Actually, Einstein himself made no such assertion (though many others have), and (depending, as Bill Clinton would say, on what you mean by "nothing") the statement is demonstrably false. Imagine shining a flashlight on a distant wall. As you swing it from left to right, the spot on the wall moves at a speed that depends on how far away the wall is, and there is no limit on how fast the spot can move. But is a spot a "thing"? It does not transport energy from one

[30] Bell's argument is simple, beautiful, and general. It astonished the scientific community – which, if you think about it, is very strange. It suggests that, after all, a lot of physicists were not entirely comfortable with the Copenhagen interpretation, and secretly expected it to be superseded by a deterministic theory. How else can one explain how shocked they were to learn that the orthodox view is right, after all?

[31] Bell's theorem and the experiments it inspired do not rule out *non*-local hidden variable theories, such as Bohm's. You can have a deterministic theory that fits the data, if you insist, but it will necessarily include spooky action-at-a-distance. That being the case, Occam's razor favors the simplest theory: ordinary quantum mechanics.

place on the wall to another, nor can it communicate any message; it cannot transmit any *causal* signal.[32] A *causal* signal can *never* propagate faster than light, according to special relativity. If it did, a moving observer would report that the effect preceded the cause, and that, as we noted in Section 2.5, leads to inescapable logical contradictions.

So the question arises: are the superluminal influences ("spooky action-at-a-distance") required by the orthodox interpretation of quantum mechanics (and confirmed Aspect's experiments) "causal"? Not in the usual sense of the word. Mary cannot send a message to Tom – she cannot *cause* his frisbee to be counterclockwise – because she does control the spin of her own frisbee. To be sure, *some* sort of influence goes from Mary to Tom – otherwise we cannot account for the perfect (anti-)correlation of the spins. But it is a strangely ethereal influence that cannot be detected by examining Tom's results alone.

Suppose we repeat the experiment many times. The results might look like this (a plus sign means clockwise, and a minus sign means counterclockwise):

Mary	Tom
+	−
+	−
−	+
+	−
−	+
−	+
−	+
+	−
⋮	⋮

Each list by itself is perfectly random. It is only when – after the fact – we compare Tom's list with Mary's that we discover the remarkable anticorrelation in the data (whenever hers is plus, his is minus, and vice versa). Observer A says Mary's measurements came first, precipitating the collapse of the wave function and forcing Tom's results; observer B (moving with respect to A) says Tom's measurements came first, collapsing the wave function and forcing Mary's results. The two stories are equally valid; there is no contradiction, since they lead to the same conclusion (perfectly anticorrelated lists of data).

---

[32] Of course, *you* (the one holding the flashlight) can send a signal to someone at the wall (by blinking the light, for example), but that signal travels at the speed of light. The question is whether the passing *spot* can carry a signal from one point (S) to another (R) on the screen. The answer is no; there's nothing a "sender" (at S) can do to manipulate the spot, so as to convey a message to the "receiver" (at R). (He could send instructions to *you*, back at the flashlight, but the resulting message having travelled to you and back, would take much longer.)

If Mary's measurements had *caused* Tom's results, in a way that she could control and Tom could recognize (by examining his list alone), then there really would be an irreconcilable conflict between quantum mechanics and special relativity. But the nonlocal influences associated with the instantaneous collapse of the wave function are not causal, and the two theories are (just barely) compatible.

3.4.5 Schrödinger's cat

Einstein was not alone in rejecting the idea of indeterminacy. In 1935, Schrödinger posed his famous **cat paradox**:[33]

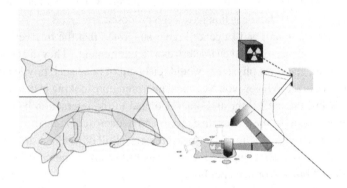

A cat is placed in a steel chamber, together with the following hellish contraption. . . . In a Geiger counter there is a tiny amount of radioactive substance, so tiny that maybe within an hour one of the atoms decays, but equally probably none of them decays. If one decays then the counter triggers and via a relay activates a little hammer which breaks a container of cyanide. If one has left this entire system for an hour, then one would say the cat is living if no atom has decayed. The first decay would have poisoned it. The wave function of the entire system would express this by containing equal parts of the living and dead cat.

At the end of the hour, the cat is neither alive nor dead, but a combination of the two . . . until a measurement occurs (say, you peek in the window). At that moment your observation forces the cat to "take a stand": dead or alive. And if you find her to be dead, then it's really *you* who killed her, by looking in the window.

Schrödinger regarded this as patent nonsense, and I think most people would agree with him. There is something absurd about the very idea of a macroscopic

---

[33] Translation by Josef M. Jauch, *Foundations of Quantum Mechanics*, Addison-Wesley, Reading (1968), p. 185.

object being in an indeterminate state. I may not *know* whether a cat is alive or dead, but in point of fact it's got to be one or the other. An electron can be in a combination of spin up and spin down, but a cat simply cannot be in a combination of alive and dead. And the notion that peeking in the window could kill the cat is surely preposterous.

The cat paradox raises two related questions:

(1) What constitutes a "measurement," in the sense of Born's statistical interpretation?
(2) How does the indeterminacy of the microscopic world become the determinacy of the macroscopic world?

Neither question has an entirely satisfactory answer, even at this late date. I'll just sketch some of the ideas that have been proposed.

John von Neumann and Eugene Wigner suggested that the intervention of human consciousness is what characterizes a "measurement." They did not offer a mechanism, and most physicists would reject out of hand such privileging of our species. (In the Wigner–von Neumann interpretation, looking in the window really *did* kill the cat.) Bohr and others pointed to the interaction between a microscopic system (the radioactive atom) and a macroscopic instrument (the Geiger counter) as the locus of the "measurement." Still others have argued that a measurement occurs when an irreversible record is left – when, for example, the glass vial breaks (or does not break).

In any event, most physicists today would agree that the collapse of the wave function occurred long before you looked in the window. The cat was never both alive and dead, because the "measurement" had already occurred when the poison was released. But it is an embarrassing fact that the concept of measurement, which is so central to quantum mechanics, is poorly understood, seriously ambiguous, and dangerously misleading.

That resolves, perhaps, this specific paradox (and absolves you of the crime), but it leaves open the question of why one cannot (apparently) produce indeterminate states of macroscopic systems. Why is a baseball, for example, definitely in Seattle or definitely in San Francisco, whereas an electron, in principle, could be in both places with equal probability? After all, macroscopic objects are made out of microscopic particles – why should they behave in such a radically different way?

This is a subject of active research and speculation, but a consensus seems to be developing. It goes like this: Every macroscopic object is constantly interacting with its environment – even in "empty" space, a baseball would be bombarded by electromagnetic radiation. It is, if you like, undergoing repeated "measurement." If you put that baseball into an indeterminate state, its wave function would quickly collapse, leaving it in a definite state. This process is called **decoherence**, and the only real question is how long it takes. Estimates have been made, and even for relatively simple objects the time it takes to "decohere" is fantastically small – minute fractions of a second. Individual electrons, by contrast, interact much less frequently with their surroundings, and they decohere far more slowly. If you could somehow shield a macroscopic object from absolutely all environmental influences, then it would remain in an indefinite state . . . indefinitely. (But you can't.)

Problem 17. For each of the following terms, give a brief (2–3 sentence) explanation, in your own words, of what it means in quantum mechanics.

(1) Statistical interpretation.
(2) Nonlocality.
(3) Entangled state.
(4) Schrödinger's cat.
(5) Decoherence.

4

Elementary particles

Elementary particle physics addresses the question, "What is matter made of?" on the most fundamental level – which is to say, on the smallest scale of size.

It's a remarkable fact that matter, at this scale, comes in tiny chunks, separated by vast empty spaces. This is radically different from our everyday experience, where matter appears to fill everything. Solids seem... well... *solid*, liquids are smooth and continuous, and even gases, such as the air in a room, occupy every nook and cranny. But the world of the very small is much more like the night sky, with pinpoints of light here and there, but mostly just nothing.

Even more surprising, these "chunks" come in a relatively small number of different types – there are protons and electrons, pi mesons and neutrinos, ... only a few dozen in all. Again, this is totally different from our macroscopic world, where there are rocks and dirt clods, bananas and chimpanzees, trees, books, and people – variety without limit.

Most astonishing of all, the chunks of any particular type are not just "pretty similar," like two Fords coming off the same assembly line, but *absolutely utterly identical*. There aren't fat electrons and skinny ones, or young electrons and old electrons, or happy electrons and sad electrons – if you've seen one, you've seen them all. There is nothing remotely like this absolute identicalness in everyday life – it is a quantum phenomenon, inconceivable in classical physics, where you could always, if necessary, paint a red spot on the object, or stamp a serial number on it. But you can't paint a spot on an electron, and if you blink, you can't even be sure that two electrons didn't change places. This utter indistinguishability underlies the Pauli exclusion principle, without which most of chemistry would be impossible.

We call these chunks of matter "elementary particles." The adjective ("elementary") is supposed to indicate that they are *not* composed of even smaller constituents. But it is a recurring theme, in this subject, that what one generation takes to be elementary, the next generation finds to be composite. In the nineteenth century, atoms were considered elementary (the very word comes from Greek, meaning "indivisible"), but as we shall see they are actually composite structures, and their constituents (protons and neutrons) are themselves composite. So the term "elementary" can be slippery, and in telling the story I will use it loosely, to denote the particles that people at the time thought were elementary.

Elementary particle physics has two tasks: (1) to identify the actors (the list of particles), and (2) to figure out how they talk to one another – how they interact. I'll approach the first part historically; it's an amazing saga, and this perspective helps you to keep track of a lot of unfamiliar players.

4.1 The early period (1897–1932)

4.1.1 Electrons, protons, and neutrons

Toward the end of the nineteenth century it was known that a hot metal filament (or "cathode") emits rays, which can be detected when they light up a nearby fluorescent screen (this is the basic element in an old-fashioned television). As the name suggests, **cathode rays** were presumed to be some sort of invisible radiation, like ultraviolet light or X-rays. But in 1897 J. J. Thomson showed that these "rays" could be deflected by electric and magnetic fields, indicating that they carry electric charge. He found that the charge was in fact *negative*, and he was able to determine the charge-to-mass ratio, e/m. Although he could not measure e or m separately, it was clear that a cathode "ray" was actually a stream of charged particles; they came to be called **electrons**.[1]

The charge of the electron was measured by Robert Millikan, in 1909. It so happens that the tiny droplets of oil produced by an "atomizer" (perfume sprayer) often pick up (or lose) a few electrons. By watching them move up or down in an electric field, Millikan was able to determine the charges of the drops. The smallest charge he got was

$$e = 1.60 \times 10^{-19} \, \text{C}, \tag{4.1}$$

but some drops had a charge of $2e$, or $3e$, or $-e$, etc. He concluded that the charge of the electron itself was $-e$, and that the droplets had lost (or gained) one, two, three, . . . of them. Combining Millikan's e with Thomson's value for

---

[1] The name itself was introduced by George Stoney, in 1894, to denote the hypothetical smallest unit of electric charge.

e/m yields the mass of the electron,

$$m_e = 9.11 \times 10^{-31} \text{ kg}. \tag{4.2}$$

Thomson guessed (correctly) that electrons are basic constituents of atoms. However, since atoms are electrically neutral, and much heavier than electrons, there had to be some other stuff, with positive charge and most of the mass. He pictured the atom as a sort of "plum pudding," with electrons embedded (like the plums) in a heavy positive dough. But Thomson's **plum pudding model** was decisively repudiated by Rutherford's scattering experiments. Rutherford and his assistants fired alpha particles (more on these in a moment) into a sheet of gold foil. Most of the alpha particles passed right through, but a few bounced right back, indicating that they had hit something small and heavy. Rutherford concluded that the atom contains a central core, the **nucleus**, which carries the positive charge and most of the mass. The much lighter, negatively charged electrons orbit around the nucleus, somewhat like planets going around the Sun, except that they are held in orbit by the electrical attraction of opposite charges instead of the gravitational attraction of two masses.

The nucleus of the lightest atom (hydrogen) was given the name **proton** by Rutherford. It carries a charge equal and opposite to that of the electron, and weighs 2000 times as much,

$$m_p = 1.673 \times 10^{-27} \text{ kg}. \tag{4.3}$$

Heavier atoms contain more protons, and more electrons. Helium has two electrons orbiting two protons, lithium has three of each, and so on. But there's a problem: helium doesn't weigh twice as much as hydrogen, but *four* times as much; lithium isn't three times hydrogen, but seven times; beryllium is nine times, ... This dilemma was finally resolved in 1932, with James Chadwick's discovery of the **neutron**, the proton's electrically neutral twin:

$$m_n = 1.675 \times 10^{-27} \text{ kg}. \tag{4.4}$$

The typical helium nucleus, it turns out, contains two neutrons, in addition to its two protons; the lithium nucleus normally has four neutrons, beryllium has five, and so on.

If you had asked a competent physicist in 1933 to name the elementary particles (or for that matter if you ask most chemists today), the answer would be protons (p), neutrons (n), and electrons (e). It's not a bad model – it explains the structure of atoms, and it provides a framework for understanding nuclei. It accounts for all of chemistry, and presumably therefore for all of biology. But the whole truth is very much richer, as we shall see.

4.1.2 Atoms

In 1869, Dmitri Mendeleev arranged the chemical elements into his famous **Periodic Table**, with the lightest atoms at the upper left, and increasing mass as you progress down, row by row, to the lower right. The vertical columns associate elements with similar chemical behavior, from the alkali metals (lithium, sodium, potassium, ...) on the left to the noble gases (helium, neon, argon, ...) on the right. At the time, there were three conspicuous holes on Mendeleev's table, but these missing elements (gallium, scandium, and germanium) were soon discovered.

Mendeleev had no idea *why* the elements fell into such a nice pattern – this was, after all, well before Thomson's discovery of the electron, Rutherford's discovery of the nucleus, and Bohr's theory of hydrogen, not to mention Schrödinger's equation and the advent of quantum mechanics. But once these essential components were assembled, everything fell into place.

The different atoms on the Periodic Table are distinguished by the number of protons in the nucleus (the **atomic number**, Z), or (what is the same thing, for a neutral atom) the number of orbiting electrons. This is the top number in each box; it is followed by the chemical symbol and the name of the element. At the bottom is the atomic mass, in **atomic mass units**, u. (One u is essentially the mass of the proton or the neutron; more precisely, 1 u = 1.6605×10^{-27} kg.)

The electrons in an atom are held in orbit by the electrical attraction of the nucleus; the potential energy of an electron is

$$V = -k\frac{Ze^2}{r}, \tag{4.5}$$

the same as hydrogen (Eq. (3.8)), except that there are now Z protons in the nucleus. The allowed energies are again characterized by the **principal quantum number** $n = 1, 2, 3, \ldots$, and the energy of an electron is given by the modified Bohr formula

$$E_n = -\frac{Z^2}{n^2}(13.6\,\mathrm{eV}). \tag{4.6}$$

The energies are negative (you would have to do work to pull the atom apart), and they increase (toward zero) as n gets larger. There are, in fact, $2n^2$ solutions to the Schrödinger equation, for a given n – that is to say, $2n^2$ different possible orbits with the same energy. For $n = 1$ (the so-called K **shell**) there are 2 allowed states; for $n = 2$ (the L **shell**) there are 8; for $n = 3$, 18; for $n = 4$, 32, and so on.

Ordinarily, any physical system tends to settle into the state of lowest energy (a marble, for example, rolls to the bottom of an empty salad bowl), unless you

The Periodic Table of the Elements

1																	2
H Hydrogen 1.00794																	**He** Helium 4.003
3 **Li** Lithium 6.941	4 **Be** Beryllium 9.012182											5 **B** Boron 10.81	6 **C** Carbon 12.0107	7 **N** Nitrogen 14.00674	8 **O** Oxygen 15.9994	9 **F** Fluorine 18.9984032	10 **Ne** Neon 20.1797
11 **Na** Sodium 22.989770	12 **Mg** Magnesium 24.3050											13 **Al** Aluminum 26.981538	14 **Si** Silicon 28.0855	15 **P** Phosphorus 30.973761	16 **S** Sulfur 32.066	17 **Cl** Chlorine 35.4527	18 **Ar** Argon 39.948
19 **K** Potassium 39.0983	20 **Ca** Calcium 40.078	21 **Sc** Scandium 44.955910	22 **Ti** Titanium 47.867	23 **V** Vanadium 50.9415	24 **Cr** Chromium 51.9961	25 **Mn** Manganese 54.938049	26 **Fe** Iron 55.845	27 **Co** Cobalt 58.933200	28 **Ni** Nickel 58.6934	29 **Cu** Copper 63.546	30 **Zn** Zinc 65.39	31 **Ga** Gallium 69.723	32 **Ge** Germanium 72.61	33 **As** Arsenic 74.92160	34 **Se** Selenium 78.96	35 **Br** Bromine 79.904	36 **Kr** Krypton 83.80
37 **Rb** Rubidium 85.4678	38 **Sr** Strontium 87.62	39 **Y** Yttrium 88.90585	40 **Zr** Zirconium 91.224	41 **Nb** Niobium 92.90638	42 **Mo** Molybdenum 95.94	43 **Tc** Technetium (98)	44 **Ru** Ruthenium 101.07	45 **Rh** Rhodium 102.90550	46 **Pd** Palladium 106.42	47 **Ag** Silver 107.8682	48 **Cd** Cadmium 112.411	49 **In** Indium 114.818	50 **Sn** Tin 118.710	51 **Sb** Antimony 121.760	52 **Te** Tellurium 127.60	53 **I** Iodine 126.90447	54 **Xe** Xenon 131.29
55 **Cs** Cesium 132.90545	56 **Ba** Barium 137.327	57 **La** Lanthanum 138.9055	72 **Hf** Hafnium 178.49	73 **Ta** Tantalum 180.9479	74 **W** Tungsten 183.84	75 **Re** Rhenium 186.207	76 **Os** Osmium 190.23	77 **Ir** Iridium 192.217	78 **Pt** Platinum 195.078	79 **Au** Gold 196.96655	80 **Hg** Mercury 200.59	81 **Tl** Thallium 204.3833	82 **Pb** Lead 207.2	83 **Bi** Bismuth 208.98038	84 **Po** Polonium (209)	85 **At** Astatine (210)	86 **Rn** Radon (222)
87 **Fr** Francium (223)	88 **Ra** Radium (226)	89 **Ac** Actinium (227)	104 **Rf** Rutherfordium (261)	105 **Db** Dubnium (262)	106 **Sg** Seaborgium (263)	107 **Bh** Bohrium (262)	108 **Hs** Hassium (265)	109 **Mt** Meitnerium (266)	110 (269)	111 (272)	112 (277)	113	114				

58 **Ce** Cerium 140.116	59 **Pr** Praseodymium 140.90765	60 **Nd** Neodymium 144.24	61 **Pm** Promethium (145)	62 **Sm** Samarium 150.36	63 **Eu** Europium 151.964	64 **Gd** Gadolinium 157.25	65 **Tb** Terbium 158.92534	66 **Dy** Dysprosium 162.50	67 **Ho** Holmium 164.93032	68 **Er** Erbium 167.26	69 **Tm** Thulium 168.93421	70 **Yb** Ytterbium 173.04	71 **Lu** Lutetium 174.967
90 **Th** Thorium 232.0381	91 **Pa** Protactinium 231.03588	92 **U** Uranium 238.0289	93 **Np** Neptunium (237)	94 **Pu** Plutonium (244)	95 **Am** Americium (243)	96 **Cm** Curium (247)	97 **Bk** Berkelium (247)	98 **Cf** Californium (251)	99 **Es** Einsteinium (252)	100 **Fm** Fermium (257)	101 **Md** Mendelevium (258)	102 **No** Nobelium (259)	103 **Lr** Lawrencium (262)

do something to stir it up. One would suppose, therefore, that all the electrons in an atom would collect in the state $n = 1$. But the **Pauli exclusion principle** dictates that *no two electrons can occupy the same state*. In hydrogen the electron normally occupies the $n = 1$ state. In helium both electrons are in the $n = 1$ state. In lithium two electrons are in the $n = 1$ state, but now the K shell is full, and the third electron is forced into one of the $n = 2$ states. As we progress across the second row of the periodic table (beryllium, boron, carbon, ...) more and more electrons populate $n = 2$ states, until we reach neon, with a total of 10 electrons (2 in the $n = 1$ shell and 8 in the $n = 2$ shell). At this point the K and L shells are both full, and the next atom (sodium) is obliged to assign an electron to $n = 3$.

If this were the whole story, the third row would comprise 18 elements, as the $n = 3$ states fill up. Actually, it has only eight. What's gone wrong? Equation (4.5) correctly represents the electrical attraction of the nucleus, but it does *not* include the repulsion of the other electrons in the atom. This doesn't matter much if there are only a few of them, but when there get to be 10 or 15, they simply cannot be ignored. The outlying electrons no longer feel the full force of the nucleus – the cloud of inner electrons effectively "shields" it. The result is that some of the $n = 4$ states are actually lower in energy than some of the $n = 3$ states, and they start to populate before the $n = 3$ shell is full. The details are complicated, and do not really concern us; suffice it to say that they are well understood by atomic physicists.

The chemical behavior of an atom is determined by its outer ("valence") electrons – the ones in unfilled shells. The inner electrons, in completed shells, are very comfortable, and do not participate in chemical reactions. The elements in the first column of the Periodic Table have a single valence electron, and they share similar chemical properties. The elements in the last column have no valence electrons, and they are chemically inert (they're called "noble" gases because they do not interact much with other atoms). And the central columns associate elements with similar outer electron structure and hence similar chemical properties.

Example 1. How many neutrons are there in a carbon nucleus?

Solution: Carbon (C) has atomic number 6, so there are six protons in the nucleus (and six electrons in orbit). But the atomic mass of carbon is 12 u, so there must also be six neutrons in the nucleus. (The electrons' contribution to the total mass is negligible.)

Problem 1. How many protons are there in an oxygen nucleus? How many electrons are there in a nitrogen atom? How many neutrons are there in a sulfur nucleus?

> **Problem 2.** How many distinct electron states are there with $n = 5$? What about $n = 6$?

> **Problem 3.** Estimate the energy of a lithium atom (remember, there are three electrons, two with $n = 1$, and one with $n = 2$).

> **Problem 4.** How many valence electrons are there in helium? How about carbon? Fluorine?

4.1.3 Nuclei

A nucleus is characterized by the number of protons (Z) and the number of neutrons (N) it contains. The former is the atomic number; it identifies the element. The latter is somewhat flexible. For low-mass nuclei the number of neutrons tends to be roughly equal to the number of protons, but heavier nuclei have more neutrons than protons:

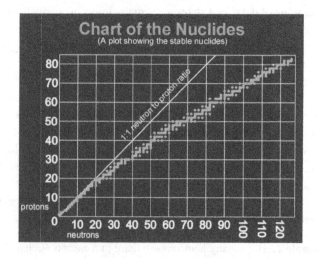

In fact, for a given atom the number of neutrons can vary. Ordinary hydrogen has no neutrons, but the rare form known as "heavy hydrogen" (or **deuterium**) has one, and **tritium** (very heavy hydrogen) has two. Most helium nuclei have two neutrons, but a few have just one. Nuclei with the same Z but different N are called **isotopes**; chemically, they are essentially identical (remember, chemical behavior involves only the outermost electrons). Specific isotopes are labeled by the total number of **nucleons** (neutrons plus protons), as a left-superscript. Thus ^{2}H would be deuterium, ^{4}He is the common isotope of helium, and ^{235}U is the uranium nucleus with 92 protons and 143 neutrons. Typically, one isotope is much more abundant than the others, and that's why most atomic masses are close to whole numbers (in atomic mass units).[2] But there are exceptions. Naturally occurring chlorine, for example, is 76% ^{35}Cl and 24% ^{37}Cl.

A **chart of the nuclides**[3] shows the known isotopes of each atom, with the proton number plotted vertically and the neutron number plotted horizontally. Rows list the isotopes of a given atom; columns list nuclei with the same number of neutrons (**isotones**); along the downward-sloping diagonals are nuclei with (approximately) the same mass (**isobars**).

If a nucleus has too few neutrons, or too many, it will be **radioactive**, and spontaneously disintegrate. Neutrons provide "padding" between the protons, which after all repel one another electrically – too few neutrons to keep the peace, and the quarreling protons will bust the nucleus apart. On the other hand, the neutron itself is intrinsically unstable (only the protective environment of a nucleus keeps neutrons from decaying) – you don't want too *many* neutrons, 'cause they're all crazy.

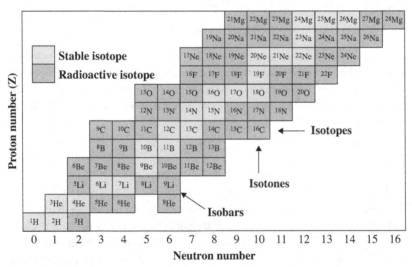

You might think that a nucleus with too many neutrons would simply spit one out, and this does occur in rare cases. Beryllium-13, for example, can emit a neutron:

$$^{13}\text{Be} \rightarrow {}^{12}\text{Be} + n$$

---

[2] The masses listed on the Periodic Table are *averages* over the naturally occurring isotopic abundances. If you want the mass of a particular isotope, you must go to a more detailed listing. By the way, the atomic mass unit (u) is defined so that the mass of a carbon-12 atom is exactly 12 u.

[3] A **nuclide** is an atom with a specific number of protons and neutrons in its nucleus.

(because the number of protons has not changed, it's still beryllium – just a different isotope). But a much more common mechanism is for a neutron to decay into a proton plus an electron.[4] For instance,

$$^{14}C \rightarrow {}^{14}N + \beta^-$$

(this time the **daughter** nucleus has one more proton, so it's a different element). In the early days it was not realized that the emitted particle is an ordinary electron; it was called a "beta particle" (β^-), and the process itself is known as **beta decay**.

Similarly, you might guess that a nucleus with too few neutrons (which is to say, too many protons) would simply eject a proton. But proton emission is extremely rare; it was first observed in cobalt-53:

$$^{53}Co \rightarrow {}^{52}Fe + p.$$

More common is a form of beta decay in which a proton converts into a neutron plus a positron (antiparticle of the electron, known in this context as β^+).[5] For example, sodium-22 goes to neon-22 plus a positron:

$$^{22}Na \rightarrow {}^{22}Ne + \beta^+.$$

For heavy nuclei with an excess of protons, a common process is **alpha decay**, in which an **alpha particle** (α) is emitted. An alpha particle is a package of two protons and two neutrons – the nucleus of ^{4}He. It turns out that protons and neutrons just *love* to get together in this combination. A typical alpha decay is uranium-238 to thorium-234:

$$^{238}U \rightarrow {}^{234}Th + \alpha$$

(notice that the mass decreases by 4, and the atomic number by 2).

There is no way to predict when a given unstable nucleus will disintegrate. The process is governed by quantum mechanics, and all it can tell you is probabilities. And remember: elementary particles have no *age*, so a nucleus is just as likely to decay in the next second, regardless of how long it has been around.[6] But if you start out with a bucketfull of radioactive atoms, all of the same kind, after a certain amount of time, called the **half-life** for that species, half of them have decayed. Wait another half-life, and half of the remainder

---

[4] Also a neutrino, but that's getting ahead of the story.

[5] Because the neutron weighs more than the proton, this could never happen in isolation, but within a nucleus the proton can "steal" the extra energy from its neighbors, which rearrange themselves into a lower-energy configuration. A closely related process is **electron capture**, in which a proton grabs an orbiting electron and undergoes **inverse beta decay**: $p + e \rightarrow n$.

[6] This is radically different from the biological paradigm. A ninety year old person is much more likely to die in the next year than a teenager, but that's not true for unstable particles. Mathematically, this "agelessness" leads to the exponential decay law.

will disintegrate, and so on. After a time t, the number *remaining* will be

$$N(t) = \left(\tfrac{1}{2}\right)^{t/\tau} N_0, \qquad (4.7)$$

where N_0 is the number you started with, and τ is the half-life. This **exponential decay law** applies to all spontaneous decay processes in nuclear and particle physics.

> **Example 2.** Suppose the half-life of some radioactive nucleus is 1 second. If you start with 100 atoms, how many would still be left after 1, 2, 3, ... , seconds?
>
> *Solution:* After 1 second half of them would remain: 50. After 2 seconds half of those would be gone, leaving 25. After 3 seconds half of *those* decay, and we are left with 12.5. (Hmmm ... What does that mean? I can't have half a nucleus! Decays, remember, are probabilistic. So, maybe 12, maybe 13 – somewhere around there.) After 4 seconds, 6.25; after 5 seconds 3.125; after 6 seconds 1.5625; after 7 seconds 0.781 25 – fewer than 1! After 8 seconds the chances are none at all would remain.

The half-lives of radioactive nuclei vary over a fantastic range, from 10^{-16} s for beryllium-8 to 10^{32} s for tellurium-128. Here's a (very incomplete) list, just to give you a sense of the variation.

Isotope	Half-life	Mode
^{8}Li	0.84 seconds	β^-
^{18}F	1.8 hours	β^+
^{57}Co	270 days	ec
^{22}Na	2.6 years	β^+
^{90}Sr	29 years	β^-
^{14}C	5700 years	β^-
^{59}Ni	76 000 years	β^+
^{10}Be	1.4 million years	β^-
^{235}U	700 million years	α
^{238}U	4.5 billion years	α

Atomic masses have been measured with great precision. The mass of a nuclide with Z protons and N neutrons is approximately $(Z + N)$u. More precisely, it will be slightly less than Z times the mass of the proton plus the mass of the electron, plus N times the mass of the neutron:

$$m = Z(m_p + m_e) + N(m_n) - \frac{\epsilon}{c^2}. \qquad (4.8)$$

Here ϵ is the **binding energy** – the amount of energy it would take to pull the nucleus apart into its constituent nucleons;[7] it contributes to the total mass via Einstein's formula, $E = mc^2$.

The binding energy per nucleon, $\epsilon/(Z + N)$, varies from nucleus to nucleus; for stable nuclei it rises steeply as Z increases (with a spike for the unusually tightly bound ^{4}He), hits a maximum around iron-56, and tapers off for the heavier elements:

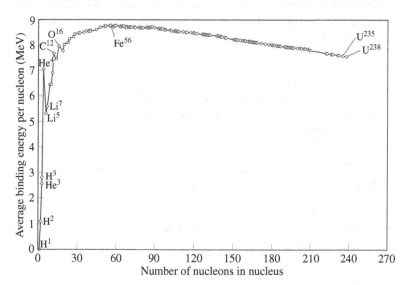

Just as water, left to its own devices, flows downhill to the lowest accessible point, neutrons and protons would "like" to be in the most tightly bound state. But, like water caught in an isolated mountain lake, most of them are (at present) stuck in other nuclei. If something comes along to shake things up, they will tend to head in the direction of iron and nickel. Heavier nuclei break apart, in a process called **fission**, and lighter nuclei join together in **fusion**.

In very heavy nuclei fission can occur spontaneously.[8] For example,

$$^{252}\text{Cf} \rightarrow {}^{140}\text{Xe} + {}^{108}\text{Ru} + 4n.$$

More often, fission is induced by the absorption of a neutron:

$$n + {}^{235}\text{U} \rightarrow {}^{236}\text{U} \rightarrow {}^{99}\text{Zr} + {}^{134}\text{Te} + 3n.$$

---

[7] In principle, the binding energy of the electrons should also be included, but it makes a negligible contribution. The binding energy per nucleon is several MeV (million electron volts), whereas the binding energy of the electron in hydrogen, for instance, is 13.6 eV – a million times smaller.

[8] Actually, alpha decay could be regarded as a kind of fission, in which one daughter nucleus is ^{4}He. But ordinarily "fission" refers to the break-up of a nucleus into two fragments of roughly equal mass.

As these examples suggest, the fission process typically releases one or more neutrons, which can in turn trigger more fissions, leading (potentially) to a **chain reaction**. Fission is the mechanism at work in nuclear power plants and the atomic bomb.

Fusion occurs when two light nuclei merge, making a heavier nucleus. For instance

$$^3\text{He} + {}^3\text{He} \rightarrow {}^4\text{He} + 2\,{}^1\text{H}.$$

Because nuclei are positively charged, they repel one another, and it is hard to squeeze them together close enough to fuse. One solution is to heat the sample up to fantastic temperatures, and let random collisions do the job. This is what happens in the Sun, where a complicated sequence of reactions produces helium from hydrogen; it is also the process at work in hydrogen bombs. Fusion would be a great source of energy, if it weren't so difficult to contain and control something that hot.

Example 3. The mass of a helium-3 atom[9] is 3.0160 u, the mass of a helium-4 atom is 4.0026 u, and the mass of a hydrogen atom is 1.0078 u. How much energy is liberated in the fusion process $^3\text{He} + {}^3\text{He} \rightarrow {}^4\text{He} + 2\,{}^1\text{H}$? [Neglect the initial kinetic energy (if any) of the ^3He nuclei.]

Solution: The initial mass was $2 \times 3.0160 = 6.0320$ u; the final (rest) mass is $4.0026 + (2 \times 1.0078) = 6.0182$ u. So the rest mass lost is 0.0138 u, or $(0.0138) \times (1.66 \times 10^{-27}) = 2.29 \times 10^{-29}$ kg. The (kinetic) energy released (which is the same as the rest energy lost, Δmc^2) is

$$2.29 \times 10^{-29}(3.00 \times 10^8)^2 = 2.06 \times 10^{-12}\,\text{J}.$$

Problem 5. How many isotopes of helium are there? (You'll need to look this up.) List the number of neutrons for each one. What about uranium?

Problem 6. The half-life of tritium is 12 years. If you started with a gram of tritium, how much would be left after 60 years? About how long would it take before it was all gone (less than one atom, on average, remaining)? The mass of tritium is 5×10^{-27} kg.

---

[9] The electrons are irrelevant to the process, and it doesn't matter whether you work with the nuclei alone, or with the atoms. The masses are slightly different, of course, since the atomic mass includes the electrons, but the *change* in mass is the same.

Problem 7. What is the daughter nucleus, in each of the following reactions?

(a) β^- decay of ^{3}H (tritium).
(b) β^- decay of ^{60}Co.
(c) β^+ decay of ^{10}C.
(d) α decay of ^{210}Po.
(e) α decay of ^{241}Am (commonly used in smoke detectors).

Problem 8. How much energy is released in the fission reaction

$$n + {}^{235}\text{U} \rightarrow {}^{236}\text{U} \rightarrow {}^{99}\text{Zr} + {}^{134}\text{Te} + 3n?$$

Assume the initial neutron is slow, so its kinetic energy can be ignored. The mass of a ^{235}U atom is 235.044 u, the mass of ^{99}Zr is 98.917 u, and the mass of ^{134}Te is 133.912 u.

Problem 9. In the Sun, four protons fuse to make an alpha particle and two positrons:[10]

$$4\,p \rightarrow \alpha + 2\,\beta^+.$$

(The mass of an alpha particle and two positrons is the same as the mass of a ^{4}He atom: 6.646×10^{-27} kg; the mass of a proton is given in Eq. (4.3).) How much energy is released in this process?

4.2 The middle ages (1930–1960)

4.2.1 Neutrinos (1930–1956)

By 1930 it was clear that something was wrong with the story of beta decay, as I have told it: a radioactive nucleus (A) decays into a lighter nucleus (B), with the emission of an electron (known in this context as a "beta particle"),

$$A \rightarrow B + e.$$

The underlying process is the decay of a neutron into a proton and an electron.

$$n \rightarrow p + e.$$

---

[10] This doesn't happen all in one step, and it also emits two neutrinos and a photon, but these details are irrelevant to the problem.

Thus, for example, the potassium isotope with 19 protons and 21 neutrons goes to calcium, with 20 of each. If the "parent" nucleus (A) is at rest, the electron flies off in one direction and the "daughter" (B) recoils in the opposite direction. The (relativistic) conservation laws determine the energy of the electron:

$$E = \left(\frac{m_A^2 - m_B^2 + m_e^2}{2m_A} \right) c^2. \qquad (4.9)$$

Every electron in this process should emerge with exactly the same energy. But in practice, the electrons have a whole *range* of energies; Eq. (4.9) tells you the *maximum* electron energy, but in most decays the electron's energy is a good deal smaller than E.

Bohr, who should have known better, was ready to abandon the law of conservation of energy. Pauli had a more inspired solution. He proposed that a third (unseen) particle was emitted in beta decay – something very light and electrically neutral. He called it the "neutron." Most people thought the idea was crazy, and in 1932 Chadwick preempted the name. But Enrico Fermi took it seriously, and in 1933 he constructed a brilliantly successful quantitative theory of beta decay by incorporating Pauli's particle, which he called the **neutrino** (ν – the Greek letter "nu"):

$$A \to B + e + \nu.$$

The underlying process (decay of the neutron) is

$$n \to p + e + \nu. \qquad (4.10)$$

The energy is now *shared* by the electron and the neutrino, and that explains why some electrons have more, and some less, up to a maximum of E (when some lucky electron gets all of it).

The neutrino is extraordinarily elusive. A typical neutrino could penetrate many light-years of lead with no significant deflection. Hundreds of billions of them pass through your thumb every second, coming from beta decays in the Sun. (They hit you from above during the day, and from below at night – having passed right through the Earth.) So it wasn't until 1956 that Pauli's conjecture was finally confirmed in the laboratory, by Frederick Reines and Clyde Cowan. They set up their detectors outside the Savannah River nuclear reactor, where enormous numbers of neutrinos made up for the fact that each one interacts so weakly.

From the beginning, it was clear that neutrinos must be extremely light – otherwise, the maximum electron energy in beta decay would be reduced by the rest energy of the neutrino. Until recently, most physicists assumed (for no good reason) that neutrinos are in fact *massless*. But by 2001, the observation of **neutrino oscillations** proved that they actually have a mass greater than zero – though to this day we do not know what that mass *is*.

Problem 10. Derive Eq. (4.9). [*Hint:* This isn't easy, so I'll walk you through it. Start with conservation of energy: $E_A = E_B + E_e$, and note that since A is at rest $E_A = m_A c^2$. Now use Eq. (2.14) to express E_B in terms of p_B, and invoke conservation of momentum to show that $p_B = -p_e$. Finally, use Eq. (2.14) again, to express p_e in terms of E_e. Solve algebraically for E_e.]

Problem 11. What is the maximum possible electron energy, in the decay of a neutron at rest? What is the *minimum* electron energy?

4.2.2 Mesons (1934–1947)

The original trinity (electrons, protons, and neutrons) was spectacularly successful in explaining atomic and nuclear physics. And yet, this model raises an obvious problem: What holds the nucleus together? After all, the positively charged protons should repel one another violently, while the neutrons, having no charge, should just wander off on their own. Evidently there must be some *other* force, more powerful than the force of electrical repulsion, that binds the protons (and neutrons) together. Physicists of that less imaginative age called it, simply, the **strong force**.

If there exists such a potent force in nature, though, why don't we notice it in everyday life? The fact is that virtually every force we experience directly, from the contraction of a muscle to the explosion of dynamite, is electromagnetic in nature; the only exception (unless you work around nuclear reactors or atomic bombs), is gravity. The answer must be that, powerful though it is, the strong force has a very short **range**. (The range of a force is like the arm's reach of a boxer – beyond that distance it can't touch you.) The range of the strong force is about the size of the nucleus itself: 10^{-15} m.

The first theory of the strong force was proposed by Hideki Yukawa in 1934. Planck and Einstein showed that electromagnetic forces are mediated by the exchange of photons[11] (the "quanta" of the electromagnetic field). Yukawa asked: What are the properties of the particle (analogous to the photon) whose

---

[11] Planck and Einstein were talking specifically about electromagnetic waves, but in quantum theory *all* fields are quantized. To the evolution of force models that I discussed at the beginning of Section 1.3 (contact → action-at-a-distance → field), quantum field theory adds a fourth: exchange of quanta. In this picture, the mutual attraction of electrons and nuclei is due to a continuous exchange of photons between them, as we'll see in Section 4.4.

exchange between protons and neutrons would account for the known features of the strong force? For example, the short range of the force indicated that the mediator would be rather heavy; Yukawa calculated that its mass should be nearly 300 times that of the electron, or about a sixth the mass of a proton. Because it fell between the electron and the proton, Yukawa's particle came to be known as the **meson** (meaning "middle-weight"). In the same spirit the electron and the neutrino are **leptons** ("light-weight"), whereas the proton and neutron are **baryons** ("heavy-weight").

Yukawa knew that no such particle had ever been observed in the laboratory, and I suppose he assumed his theory was wrong. But at the time a number of systematic studies of **cosmic rays**[12] were in progress, and by 1937 two different groups (Anderson and Neddermeyer on the West Coast, and Street and Stevenson on the East Coast) had identified particles matching Yukawa's description. Indeed, the cosmic rays with which you are being bombarded every few seconds as you read this consist primarily of just such middle-weight particles.

For a while everything seemed to be in order. But as more detailed studies of the cosmic ray particles were undertaken, disturbing discrepancies began to appear. Finally, in 1946, experiments in Rome demonstrated that the cosmic ray particles interact very weakly with atomic nuclei. If this was really Yukawa's meson, the transmitter of the strong force, the interaction should have been huge.

The puzzle was resolved in 1947, when Cecil Powell and his co-workers at Bristol discovered that there are actually *two* middle-weight particles in cosmic rays, which they called π (or "pion") and μ (or "muon"). The true Yukawa meson is the π; it is produced copiously in the upper atmosphere, but disintegrates long before reaching the ground, so Powell's group put their detectors on mountaintops. One of the decay products is the lighter (and longer-lived) μ, and it is primarily muons that one observes at sea level. In the search for Yukawa's meson, the muon was simply an impostor, having nothing to do with the strong interactions. In fact, it behaves in every way like a heavier version of the electron, and properly belongs in the *lepton* family.

The figure below shows one of Powell's photographic plates (first published in 1949). A pion comes in at the lower right, and decays into a muon plus a

---

[12] Cosmic rays are due to streams of particles – mostly protons – coming in from outer space. They strike atoms in the upper atmosphere, causing cascades of secondary particles that rain down on us all the time. The ultimate source of cosmic rays is still something of a mystery, but the energy of these particles can be fantastic – far greater than anything we could produce in the laboratory.

neutrino:

$$\pi \rightarrow \mu + \nu. \tag{4.11}$$

The muon heads upward (in the picture), but the neutrino leaves no track.[13] Meanwhile, at the top of the figure the muon decays into an electron and two neutrinos (which are again invisible):

$$\mu \rightarrow e + 2\nu. \tag{4.12}$$

You'll notice that the neutrino, which was originally associated with beta decay (Eq. (4.10)), has now appeared in two ostensibly unrelated processes: the decay of the pion (Eq. (4.11)), and the decay of the muon (Eq. (4.12)). For a while it was simply *assumed* that it's all the same particle, but in 1962 experiments showed that there are actually two distinct neutrinos – one, ν_e, associated with the electron (in beta decay), and one, ν_μ, associated with the muon (in the decay of the pion). The decay of the muon actually involves one of each! All of this makes much better sense now than it did at the time..., but I'm getting ahead of the story.

---

[13] When *charged* particles pass through a photographic emulsion – or, later, a **cloud chamber**, a **bubble chamber**, or a **spark chamber** – they ionize nearby atoms, leaving a trail of spots when the film is developed (or droplets, or bubbles, or sparks, as the case may be). That's how we make visible the trajectories of charged particles. But neutral particles don't ionize atoms – they don't produce the electric fields necessary to expel electrons – so they leave no tracks. That's one of the reasons why neutrinos were so devilishly difficult to detect.

Problem 10. Explain why Powell had to go to the top of a high mountain to detect pions, whereas muons are easily detected at sea level. [*Hint:* Refer to Problem 7 in Chapter 2. The lifetime of a pion is about 1/100 the lifetime of a muon.]

Problem 11. Muons have the same electric charge as electrons, but 200 times the mass. If you could make "muonic hydrogen," with a muon in place of the electron (it's been done, by the way), what would its "Bohr radius" be (Eq. (3.13))? What would its ground state energy be (Eq. (3.15))?

4.2.3 Strange particles (1947–1960)

For a brief period in 1947, it was possible to imagine that the major problems of elementary particle physics had been solved. After a lengthy detour in pursuit of the muon, Yukawa's meson (the π) had finally been discovered. The purpose and role of the muon was something of a mystery ("Who ordered *that?*" Isidor Rabi asked). On the whole, though, it looked in 1947 as if the job of elementary particle physics was essentially done.

But this state of affairs did not last long. Later that same year, as physicists continued the exploration of cosmic rays, new and completely unexpected particles appeared. It is some measure of the surprise they elicited that these newcomers were called **strange particle**. Some of them (the Λ, the Σs, and the Ξs) behaved like heavier versions of the proton and neutron, so they were added to the baryon family; others (the four Ks, for example) behaved more like heavier versions of the pion – they were added to the meson family. In 1952, the first of the modern particle accelerators (the "Cosmotron," at Brookhaven) began operating; soon it was possible to produce strange particles in the laboratory (before this the only source had been cosmic rays), and the proliferation accelerated. Willis Lamb began his Nobel Prize acceptance speech in 1955 with the following words.

When the Nobel Prizes were first awarded in 1901, physicists knew something of just two objects which are now called "elementary particles": the electron and the proton. A deluge of other "elementary" particles appeared after 1930; neutron, neutrino, μ meson [*sic*], π meson, heavier mesons, and various hyperons [strange baryons]. I have heard it said that "the finder of a new elementary particle used to be rewarded by a Nobel Prize, but such a discovery now ought to be punished by a $10,000 fine."

One striking feature of the strange particles is that they are always produced in pairs. In the collision of a pion and a proton, for example, you might get[14]

$$\pi^- + p \rightarrow K^+ + \Sigma^-,$$

or

$$\pi^- + p \rightarrow K^0 + \Sigma^0,$$

or

$$\pi^- + p \rightarrow K^0 + \Lambda,$$

(in each case producing two strange particles), but you never get

$$\pi^- + p \nrightarrow \pi^+ + \Sigma^-,$$

or

$$\pi^- + p \nrightarrow \pi^0 + \Lambda,$$

or

$$\pi^- + p \nrightarrow K^0 + n,$$

(with only one strange particle in the final state). This reminded Murray Gell-Mann of the conservation of electric charge: from neutral particles you can create two (oppositely) charged particles, but never just one:

$$\pi^0 + n \rightarrow \pi^- + p,$$

but never

$$\pi^0 + n \nrightarrow \pi^0 + p.$$

He assigned a **strangeness** to each particle (zero for the "ordinary" particles, like the proton, neutron, and pion, but $+1$ for the K^+ and the K^0, -1 for the Σs and the Λ, etc.), and announced a law of **conservation of strangeness** to "explain" why you could only produce strange particles in pairs (with equal and opposite strangeness). It sounds almost ridiculous, but it turned out to be a critically important idea.

---

[14] Many particles come with a variety of different electric charges, which we indicate with a superscript (π^+, π^-, π^0, K^+, K^0, Σ^+, Σ^-, Δ^{++}, etc.), in units of the proton charge. In the case of very familiar particles (such as the proton or the electron) the charge is usually not indicated – you're just supposed to *know* that p is p^+, and e means e^-. Otherwise, if no superscript is attached, the particle is neutral (Λ is Λ^0), or the reference is generic (π for π^+, π^0, or π^-).

The garden, which seemed so tidy in 1947, had grown into a tangled jungle by 1960, and elementary particle physics could only be described as chaos. There were three families of particles: the leptons (electron, muon, and their respective neutrinos), the mesons (pions, Ks, ρs, η, ϕ, ω, . . .), and the baryons (p, n, Λ, Σs, Ξs, Δs, . . .), but there was no rhyme or reason to it all. This predicament reminded many physicists of the situation in chemistry a century earlier, in the days before the Mendeleev's Periodic Table, when scores of elements had been identified, but there was no underlying order or system. In 1960 the elementary particles awaited their own "Periodic Table."

Leptons	Mesons	Baryons
e, μ, ν_e, ν_μ	πs, Ks, ρs, η, ϕ, ω, . . .	p, n, Λ, Σs, Ξs, Δs, . . .

Problem 12. To which family (leptons, mesons, or baryons) does each of the following particles belong: (a) electron, (b) proton, (c) neutron, (d) muon, (e) pion, (f) K, (g) Λ?

Problem 13. (a) Would it be possible for a proton to decay into an electron plus a photon ($p \to e + \gamma$)? If not, why not? (b) Is this a possible reaction: $\pi^0 + n \to \pi^- + p$? (c) What about $\pi^0 + n \to \pi^+ + p$?

Problem 14. What is the strangeness of each of the following particles: (a) p, (b) Λ, (c) K^+, (d) π^-, (e) Σ^0, (f) n?

Problem 15. If the reaction $\pi^- + \Sigma^+ \to K^+ + \Xi^-$ is observed to occur, what can you conclude about the strangeness of the Ξ^-?

4.3 The modern era (1961–1978)

4.3.1 The Eightfold Way (1961)

The Mendeleev of elementary particle physics was Gell-Mann, who introduced the so-called **Eightfold Way** in 1961. The Eightfold Way assigns the baryons and mesons[15] to weird geometrical patterns, according to their charge and

---

[15] The Eightfold Way says nothing about the leptons; we'll see later on how they fit into the scheme of things.

strangeness. The eight lightest baryons fit into a hexagonal array, with two particles at the center:

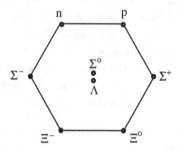

This group is known as the **baryon octet**. Notice that particles of like charge lie along the downward-sloping diagonal lines: $Q = +1$ (in units of the proton charge) for the proton and the Σ^+; $Q = 0$ for the neutron, the Λ, the Σ^0, and the Ξ^0; $Q = -1$ for the Σ^- and the Ξ^-. Horizontal lines associate particles of like strangeness: $S = 0$ for the proton and neutron, $S = -1$ for the middle row and $S = -2$ for the two Ξs.

The eight lightest mesons fill out a similar hexagonal pattern – the **meson octet**:

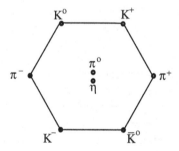

Once again, diagonal lines determine charge, and horizontals determine strangeness; but this time the top row has $S = 1$, the middle row $S = 0$, and the bottom row $S = -1$. (This discrepancy is a historical accident; Gell-Mann could just as well have assigned $S = 1$ to the proton and neutron, $S = 0$ to the Σs and the Λ, and $S = -1$ to the Ξs. In 1953 he had no reason to prefer that choice, and it seemed most natural to give the familiar particles – proton, neutron, and pion – a strangeness of zero.)

Hexagons were not the only shapes allowed by the Eightfold Way; there was also, for example, a triangular array, incorporating 10 heavier baryons – the **baryon decuplet**:

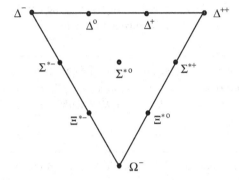

Now, as Gell-Mann was fitting particles into the decuplet, an absolutely lovely thing happened. Nine of the particles were already known experimentally, but at that time the tenth particle (the one at the very bottom, with a charge of −1 and strangeness −3) was missing: no particle with these properties had ever been detected in the laboratory. Gell-Mann boldly announced that such a particle should exist, and he told the experimentalists exactly how to produce it, how to detect it, what its mass would be, how long it would last, . . . , a whole laundry-list of properties, deduced using nothing put pencil and paper. And sure enough, in 1964 the famous **omega-minus** particle was discovered, precisely as Gell-Mann had predicted.[16]

Since the discovery of the omega-minus (Ω^-), no one has seriously doubted that the Eightfold Way is correct. Over the next 10 years, every new baryon and every new meson found a home in one of these Eightfold Way **supermultiplets**.[17]

For every particle there is an associated **antiparticle**,[18] with the same mass, but the opposite charge and strangeness. So in addition to the baryon octet (for example), there is an *anti*-baryon octet, with the (negatively charged) antiproton, the antineutron, the antilambda, and so on (antiparticles are designated by an overbar, so the antiproton, for example, would be written $\bar{p}$, the antineutron $\bar{n}$, etc.). I won't bother to draw the antiparticle supermultiplets; you can figure out the pattern for yourself by simply reversing the signs of Q and S. In the case of the mesons, there is no second diagram for the antiparticles – they

---

[16] Recall that Mendeleev, too, had confronted gaps on his table. In both cases what might at first have seemed an awkward defect in the model turned out to be a huge triumph.

[17] To be sure, there were occasional false alarms – particles that did not seem to fit Gell-Mann's scheme – but they always turned out to be experimental errors. Elementary particles have a way of appearing and then *dis*appearing. Of 26 mesons listed on a standard table in 1963, 19 were later found to be spurious!

[18] Recall that this is a deep consequence of the union of special relativity and quantum mechanics.

are already included. Thus, the anti-π^+ is the π^-, the anti-K^+ is the K^-, the anti-K^0 is the $\bar{K}^0$, and the π^0 is its *own* antiparticle.

Problem 16. There is no meson decuplet. The next-heavier mesons (known as the "vector mesons") form another octet (ρ^+, ρ^0, ρ^-, and ω with $S = 0$; K^{*+} and K^{*0} with $S = 1$; K^{*-} and $\bar{K}^{*0}$ with $S = -1$). Construct the vector meson octet.

Problem 17. (a) Construct the Eightfold Way diagram for the antibaryon octet. (Strangeness should decrease as you go down the diagram, so $\bar{p}$ and $\bar{n}$ are in the bottom row. Charge increases toward the upper right.) (b) Do the same for the antibaryon decuplet.

4.3.2 The quark model (1964)

Why do the baryons and mesons fit into these bizarre patterns? The answer was suggested already in 1964, when Gell-Mann and Zweig independently proposed that all these particles are composed of even more elementary constituents, which Gell-Mann called **quarks**. The quarks come in three types (or **flavors**): u ("up"), d ("down"), and s ("strange") – and of course there are three matching *anti*-quarks:

QUARKS			ANTIQUARKS		
q	Q	S	$\bar{q}$	Q	S
u	$+2/3$	0	$\bar{u}$	$-2/3$	0
d	$-1/3$	0	$\bar{d}$	$+1/3$	0
s	$-1/3$	-1	$\bar{s}$	$+1/3$	1

There are two rules of composition.

- Every baryon is composed of three quarks (every *anti*-baryon is composed of three *anti*-quarks).
- Every meson is composed of a quark and an antiquark.

That's all there is to the quark model. It is now a matter of simple arithmetic to construct the possible baryons and mesons. All we need to do is list the combinations of three quarks (or quark–antiquark pairs), and add up their charge and strangeness.

BARYONS				
qqq	Q	S	decuplet	octet
uuu	$+2$	0	Δ^{++}	–
uud	$+1$	0	Δ^{+}	p
udd	0	0	Δ^{0}	n
ddd	-1	0	Δ^{-}	–
uus	$+1$	-1	Σ^{*+}	Σ^{+}
uds	0	-1	Σ^{*0}	Σ^{0}, Λ
dds	-1	-1	Σ^{*-}	Σ^{-}
uss	0	-2	Ξ^{*0}	Ξ^{0}
dss	-1	-2	Ξ^{*-}	Ξ^{-}
sss	-1	-3	Ω^{-}	–

MESONS				
$q\bar{q}$	Q	S	nonet	nonet
$u\bar{u}$	0	0	π^{0}	ρ^{0}
$u\bar{d}$	$+1$	0	π^{+}	ρ^{+}
$d\bar{u}$	-1	0	π^{-}	ρ^{-}
$d\bar{d}$	0	0	η	ω
$u\bar{s}$	$+1$	$+1$	K^{+}	K^{*+}
$d\bar{s}$	0	$+1$	K^{0}	K^{*0}
$s\bar{u}$	-1	-1	K^{-}	K^{*-}
$s\bar{d}$	0	-1	$\bar{K}^{0}$	$\bar{K}^{*0}$
$s\bar{s}$	0	0	η'	ϕ

There are 10 possible combinations of three quarks – perfect for the baryon decuplet, but too many for the octet. For reasons having to do with the Pauli exclusion principle (which applies to quarks as well as electrons), the three combinations that involve the *same* quark (uuu, ddd, and sss) do not occur in the octet configuration, and there are *two* particles (Σ^0 and Λ) that have one of each (uds). Meanwhile, there are *nine* meson combinations, which would appear to be one too many, but it turns out there was a ninth meson already on the books – the η', which had been classified as a "singlet" in the Eightfold Way, but now finds a home in the ("pseudoscalar") **meson nonet**.[19] (The same thing happens for the heavier "vector" meson nonet, with the singlet ϕ joining the other eight.)

Notice that the same quark combination can make more than one particle: uud yields the proton, but also the Δ^{+} (and many more baryons in other decuplets and octets); $u\bar{d}$ makes the π^{+}, but also the ρ^{+} (and many heavier mesons). Evidently there are different ways to "package" the same quarks together. This is reminiscent of the hydrogen atom – a bound state of an electron and a proton – which admits a plethora of excited states.[20] By the way, the quark model neatly explains why the antimesons reside in the same supermultiplet as the mesons: the antiparticle of the π^{+} ($u\bar{d}$), for example,

[19] You'll notice that there are three different quark/antiquark combinations with $Q = 0$ and $S = 0$, and it's not clear which is the π^0, which the η, and which the η'. Actually, they are quantum mixtures. The π^0, for example, is a 50–50 mix of $u\bar{u}$ and $d\bar{d}$, while the η is a bit of all three. But for simplicity, on the chart, I have pretended they are "pure" combinations.

[20] If we were starting over, we might well call the Δ^{+} an "excited state" of the proton, and the ρ an excited state of the π, but since they were regarded as distinct particles long before the quark model revealed that they were made out of the same constituents, I guess they get to keep their separate identities.

would be the $d\bar{u}$ (π^-). What could be simpler? The whole chaotic jungle of baryons and mesons is reduced to combinations of just three quarks!

The quark model did, however, suffer from one near-fatal embarrassment: in spite of the most diligent search, no one has ever found an individual quark. If a proton is really made out of three quarks, you'd think that when you hit it hard enough, the quarks ought to come popping out. But it doesn't seem to happen; the impact creates (for example) a $d/\bar{d}$ pair, and you're left with a neutron in one hand and a pion in the other, but no quark in sight. The failure of experiments to produce "free" quarks led to widespread skepticism about the quark model in the late 1960s and early 1970s. Those who clung to the model tried to conceal their disappointment by introducing the doctrine of **quark confinement**: quarks are *absolutely confined* within baryons and mesons, and no matter how hard you try, you just cannot get them out.[21] Of course, this doesn't explain anything; it just gives a name to our frustration.[22]

In 1964, O. W. Greenberg suggested that quarks not only come in three flavors (up, down, and strange), but each one also comes in three different **colors** – red, green, and blue, say.[23] The antiquarks have *negative* color (the antiparticle of a red u quark – with a "redness" of $+1$ – would be a $\bar{u}$ with a redness of -1). In a baryon there is one quark of each color; the combination is taken to be "colorless," mimicking the fact that mixing three primary colors is supposed to yield white. In a meson the quark and the antiquark have opposite colors (if the quark is red, the antiquark is antired), and this too is considered "colorless." Quark confinement can then be stated as follows:

Every naturally occurring particle is colorless.

Greenberg had other reasons for introducing the color hypothesis, but many people thought the whole idea was so contrived as to represent perhaps the last gasp of the quark model. As it turned out, color is the essential ingredient in the modern theory of the strong force (which is known, for that reason, as **chromodynamics**).

---

[21] Confinement does not mean quarks are inaccessible to experimental study. Around 1970 a series of "deep inelastic scattering" experiments were performed at the Stanford Linear Accelerator Center (**SLAC**), in which electrons were fired into protons, and – in an echo of the Rutherford experiment – revealed that the proton contains three lumps (called "partons" by those who did not dare to call them quarks).

[22] A really clean explanation of the mechanism responsible for quark confinement would surely win the Nobel Prize (there are fairly persuasive theories out there, but they tend to be rather cumbersome). It may even be that quark confinement is not absolute, and free quarks did exist in the first moments after the Big Bang.

[23] Of course, this has nothing to do with color in the usual sense – it's just a name for the three types. We could as well call them "type a," "type b," and "type c."

4.3.3 The November revolution (1974)

What rescued the quark model was something almost entirely unexpected: the discovery of the psi (ψ) meson, on the weekend of November 10, 1974. The ψ is an electrically neutral, extremely heavy meson – more than three times the weight of a proton (the original notion that mesons are "middle-weight" and baryons "heavy-weight" had long since gone by the boards). But what made this particle so unusual was its extraordinarily long lifetime: the ψ lasts fully 10^{-20} seconds before disintegrating. Now, 10^{-20} seconds may not impress you as a particularly long time, but you must understand that the typical lifetimes for mesons in this mass range are on the order of 10^{-23} seconds. So the ψ has a lifetime about a thousand times longer than any comparable particle. It's as though someone came upon an isolated village in Peru or the Caucasus where people live to be 70 000 years old. That wouldn't just be some actuarial anomaly, it would be a sign of fundamentally new biology at work. And so it was with the ψ: its long lifetime, to those who understood, spoke of fundamentally new physics.

In the months that followed, the true nature of the ψ meson was the subject of lively debate, but the explanation that won the day was provided by the quark model: the ψ is a bound state of a new (fourth) quark, the c (for **charm**), and its antiquark, $\psi = c\bar{c}$. Actually, the idea of a fourth flavor, and even the whimsical name, had been introduced years earlier, by Bjorken and Glashow. They noticed an intriguing parallel between the leptons and the quarks:

$$Leptons: e, \nu_e, \mu, \nu_\mu$$
$$Quarks: d, u, s$$

If all mesons and baryons are made out of quarks, these two families are the truly fundamental particles. But why *four* leptons and only *three* quarks? Wouldn't it be nicer if there were four of each?

So when the ψ was discovered, the quark model was ready and waiting with an explanation. Moreover, it was an explanation pregnant with implications. For if a fourth quark exists, there should be all kinds of new baryons and mesons, carrying various amounts of charm. As more and more of these charmed particles were discovered in the laboratory, the interpretation of the ψ as $c\bar{c}$ was established beyond any reasonable doubt. More important, the quark model itself was put back on its feet.

However, the story does not end there, for in 1975 a new *lepton* was discovered, spoiling Glashow's symmetry. This new particle (the tau) has its own neutrino, so we are up to six leptons, and only four quarks. But don't dispair, because two years later a new heavy meson (the Υ – Greek upsilon) was discovered, and quickly identified as the carrier of a fifth quark, b (for **beauty**, or **bottom**, according on your taste): $\Upsilon = b\bar{b}$.

At this point it doesn't take a genius to predict that a sixth quark (t, for **truth**, of course, or **top**) would soon be found, restoring Glashow's symmetry with six quarks and six leptons. But the top quark turned out to be extraordinarily heavy and frustratingly elusive (it weighs nearly 200 times as much as a proton); its existence was not definitively established until 1995.[24]

Problem 18. Using four quarks (u, d, s, and c), construct a table of all possible baryon configurations. For each one, list its electric charge, strangeness, and charm. (The only quark with nonzero strangeness is s, with a strangeness $S = -1$; the only quark with nonzero charm is c, with a charm $C = +1$. Don't ask me why one of them is negative and the other positive – it's just the convention.) How many baryon combinations are there in all?

Problem 19. Using four quarks (u, d, s, c, and the associated antiquarks), construct a table of all possible meson combinations. For each one, list its electric charge, strangeness, and charm. (Like charge and strangeness, the charm of an antiparticle is opposite to that of the particle.) How many of them are there in all?

Problem 20. How many baryon combinations are there using all six quarks? How many meson combinations?

Problem 21. The *net* charm of the ψ meson is zero ($+1$ for the c and -1 for the $\bar{c}$). The first baryon with "naked" charm was discovered in 1975: the Λ_c^+, with $C = 1$ and $S = 0$. Later on, others were found, including the Ω_c, with $C = 1$ and $S = -2$. The first charmed meson, the D^+ with $C = 1$ and $S = 0$, was discovered in 1976. *Question:* What is the quark content of each of these particles?

In the same sense, the net beauty of the Υ meson is zero (-1 for the b and $+1$ for the $\bar{b}$). The first baryon observed with "naked" beauty (or "bare"

---

[24] The heavier the particle, the more energy required to produce it. (In a collision, kinetic energy is converted into rest energy, so you need more of the former to generate more of the latter.) That's why elementary particles tend to be discovered in order of increasing mass. The top quark was finally created at Fermilab's **Tevatron** – before that, there simply was no machine on Earth powerful enough to do the job.

bottom)[25] was the Λ_b^0, discovered in the 1980s, about the same time as the first beautiful mesons, B^- and $\bar{B}^0$. All of these particles have $S = 0$, $C = 0$, $B = -1$; what is their quark content?[26]

4.3.4 The Standard Model (1978)

By around 1978, the modern picture of elementary particles – known prosaically as the **Standard Model** – was essentially complete. The top quark would not be detected in the laboratory for nearly two decades, but nobody seriously doubted its existence. Many baryons and mesons remained (and remain) to be discovered, but as they have appeared, over time, they have all fitbed the quark model predictions. And the neutrinos still had some surprises in store (they are not massless after all, though in retrospect even this seems perfectly natural).

According to the Standard Model, there are two basic families of elementary particles: quarks and leptons. There are six of each, and they fall into three **generations**.

	LEPTONS			QUARKS		
	l	Q	mc^2	q	Q	mc^2
First generation	e	-1	0.511	d	$-1/3$	~ 2
	ν_e	0	$< 2 \times 10^{-6}$	u	$+2/3$	~ 5
Second generation	μ	-1	106	s	$-1/3$	~ 100
	ν_μ	0	< 0.2	c	$+2/3$	1200
Third generation	τ	-1	1777	b	$-1/3$	4200
	ν_τ	0	< 18	t	$+2/3$	$174\,000$

In the third column I have listed the particle's mass (or rather, its rest energy, mc^2), in MeV (millions of electron volts). The neutrino masses are not yet established – all we can say for sure is that they are not zero, and they are extremely small (the electron, for example, weighs at least a quarter of a million times as much as the electron neutrino). The masses of the light quarks (d, u, and s) are very difficult to measure – hence the tilde (meaning "approximately"). Notice that these numbers extend over a *fantastic* range, from less than 10^{-6} to 10^5. We have absolutely no idea why the fundamental particles have these

---

[25] The silly terminology, which is by this point completely out of hand, is a reminder of how squeamish people were about taking the quark model seriously, in the early days.

[26] Incidentally, the $B^0/\bar{B}^0$ system has turned out to be extraordinarily rich, and so-called "B factories" were constructed at SLAC (called, inevitably, "BaBar") and in Japan.

particular masses – that is something for the next generation of physicists to figure out.

The second and third generations recapitulate the first, but at higher mass. You're probably wondering whether there could be more generations, even heavier than these. The answer is, (almost) unequivocally, no: there are three, and only three (though *why* this is true remains a mystery). It is possible, both from astrophysical evidence and in the laboratory, to count the number of different types of light neutrinos, and that number is 3. If there is a fourth generation, then, its neutrino will have to be very heavy. That's not impossible, I suppose, but it is contrary to all our experience.

The Standard Model has been extraordinarily successful. Indeed, for 35 years every confirmed experiment has been consistent with this simple picture. But how come – with the sole exception of the electron – we never encounter these quarks and leptons (and the associated antiquarks and antileptons) in everyday life? That's an interesting question, and it has four different answers, depending on which particle you are talking about.

- **Neutrinos.** Remember, neutrinos interact very weakly with matter. The fact is, we are absolutely *bathed* in neutrinos, but we are unaware of it because they pass right through practically anything.
- **Antiparticles.** When a charged particle meets its antiparticle, they annihilate, producing two photons. For example, if a positron encounters an electron,

$$e^+ + e^- \rightarrow \gamma + \gamma \tag{4.13}$$

(γ – Greek gamma – is the symbol for a photon). Now, our world is full of electrons; if a positron showed up in this room, it would quickly find an electron, and instantly they would both disappear. So antimatter doesn't stick around long.[27]

This raises an intriguing question: why is there so much matter around (as opposed to antimatter)? Everybody assumes that the Big Bang created equal amounts of matter and antimatter. If that is the case, then either there is a lot of antimatter hiding out there somewhere (a possibility for which there is absolutely no evidence), or else there is some mechanism that favors matter over antimatter in the course of cosmic evolution. There are speculative theories to account for this "matter/antimatter asymmetry," but for the moment it remains a tantalizing mystery.

---

[27] How can a meson state like $u\bar{u}$ survive? Why don't the quark and antiquark simply annihilate? *Answer:* they *do*! That's why the π^0 has a much shorter lifetime than its siblings, the π^+ and the π^-, to which, in other respects, it is very similar. The "decay" ($\pi^0 \rightarrow \gamma + \gamma$) is really just the annihilation of the u and the $\bar{u}$. The $\pi^+ = u\bar{d}$ cannot decay this way, because the $\bar{d}$ is not the antiparticle of the u.

- **Unstable particles.** Most elementary particles are intrinsically unstable – they disintegrate spontaneously into lighter particles, unless some conservation law prevents them from doing so. The muon, for example, decays in about a millionth of a second, the tau in 10^{-13} seconds. But the electron is stuck: there is no lighter charged particle for it to decay into. That's why we encounter lots of electrons, but very few muons and taus.
- **Quarks.** As we have seen, individual quarks do not show their face in public – they come only in groups of three (baryons) and two (mesons). The lightest baryons are the proton and the neutron – they are the stable ones;[28] the lightest mesons are the pions – that's why they are more "familiar" (and last longer) than their heavier cousins.

The world of everyday life is indeed made of protons, neutrons, and electrons. If you want to see the more exotic particles, you have to examine cosmic rays, or go to a high-energy particle accelerator, or look back to the Big Bang.

Problem 22. In 2007 the first baryon with a quark from each of the three generations was discovered, the Ξ_b^-. What is its quark content? (Note that its charge is -1.) What are its strangeness (S), its charm (C), and its beauty (B)?

Problem 23. All of the following particles were once believed to be "elementary." Some of them are now known to be composite. In each case state whether, according to the Standard Model, the particle is elementary or composite, and, if the latter, list its constituents: (a) proton, (b) neutron, (c) electron, (d) π^+, (e) muon, (f) neutrino, (g) Ω^-.

Problem 24. In the annihilation of an electron/positron pair (Eq. (4.13)), what is the energy of each photon, in MeV? Assume the original electron and positron were at rest.

Problem 25. Why can't a particle/antiparticle pair annihilate to produce just *one* photon (for instance, $e^+ + e^- \to \gamma$)? [*Hint:* Examine the process in the reference frame[29] in which they approach one another symmetrically, from opposite directions. What would the momentum of the photon be?]

[28] Actually, the free neutron is not stable – it has a half-life of about 15 minutes. But by elementary particle standards this is an astronomically long lifetime.
[29] Remember, special relativity says you are free to analyze a process from any inertial reference system; if it can't happen in one frame, it can't happen in *any* frame.

> **Problem 26.** It has been suggested that all quarks and leptons are composed
> of two even more elementary particles: c (with an electric charge of $-1/3$),
> and n (with charge 0) – together with their antiparticles, $\bar{c}$ and $\bar{n}$). You're
> allowed to combine them in groups of three particles or three antiparticles
> (*ccn*, for example, or $\bar{n}\bar{n}\bar{n}$). In this way, construct all of the eight quarks
> and leptons – including antiparticles – in the first generation. (The other
> generations are supposed to be "excited states," with the same constituents.)

4.4 Interactions

I now turn to the second question of elementary particle physics: how do
these quarks and leptons talk to each other – what are their *interactions*? As
far as we know, there are just four ways – four fundamental forces: strong,
electromagnetic, weak, and gravitational. Each of these forces is "mediated"
by a particle – the quantum of the corresponding field. **Gluons** (there are eight
of them) mediate the strong force,[30] the photon mediates electromagnetic force,
the $W^{\pm}$ and the Z mediate the weak force,[31] and the hypothetical graviton is
supposed to mediate the gravitational force. (Actually, gravity is so pitifully
weak, compared to the others, that it plays no known role in particle physics.
For that reason – and because to this day there is no fully consistent quantum
theory of gravity – I will not discuss it here. However, gravity dominates the
large-scale structure of the Universe, as we'll see in the next chapter.)

Force	Function	Mediator	mc^2 (MeV)
Strong	Holds quarks together to make baryons/mesons (holds protons/neutrons together to make nuclei)	gluon (g) (8)	0
Electromagnetic	Holds atoms together (responsible for all of chemistry)	photon (γ)	0
Weak	Causes particle decays (neutrino interactions)	$W^{\pm}$ Z	80 385 91 188
Gravity	Holds Solar System together (determines large-scale structure of the Universe)	graviton	0

[30] Yukawa's old idea that pions mediate the strong forces has long since been superseded.
[31] The Ws and the Z were anticipated on theoretical grounds since the early 1960s, but they were
not confirmed in the laboratory (at **CERN**, in Geneva) until 1983.

Richard Feynman invented a lovely diagrammatic method for visualizing particle interactions. I'll begin with electrodynamics, because that is the simplest and the best understood.

4.4.1 Electrodynamics

All electromagnetic interactions are due to the following fundamental process.

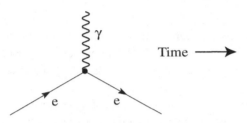

Here time flows to the right; this diagram says, "an electron (or some other charged particle) comes in, emits or absorbs a photon (I need not say which), and exits."

Imagine that I have a barrel full of these toys, made out of flexible plastic. I can snap them together, either by connecting two photon lines or by joining an electron line to another electron line (but in that case I must be careful to preserve the direction of the arrows – otherwise the snap just doesn't work). For instance, I might snap two photons together, like this.

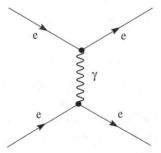

In this process two electrons came in, they exchanged a photon,[32] and they exited: $e + e \to e + e$. This is electron–electron scattering, or what in the classical theory we would interpret as the electric repulsion of two like charges. We say that the interaction was "mediated" by the exchange of a photon.

What if I rotate that diagram by 90°?

---

[32] A particle (the photon, in this case) that is internal to the diagram, is called a **virtual particle**; only the external lines (the electrons, in this case) represent "real" particles. In Problem 25 you showed that $e^+ + e^- \to \gamma$ is impossible for a real photon, but that argument does not hold for virtual particles, which do not carry their "correct" mass.

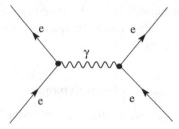

An arrow pointing "backward in time" (toward the left) is to be interpreted as the antiparticle. Evidently this diagram represents the process $e^- + e^+ \rightarrow e^- + e^+$, electron–positron scattering – or, classically, the attraction of opposite charges.

But this is not the only diagram representing electron–positron scattering. Here's another – obtained by twisting the upper half of the electron–electron diagram:

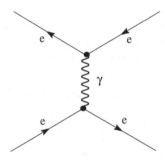

The last two diagrams describe the same physical process, but the stories they tell about *how* the process occurred are entirely different. In the first, an electron and a positron annihilated, producing a photon, which subsequently rematerialized as an electron–positron pair; in the second an electron and a positron exchanged a photon.

Now, each of these **Feynman diagrams** stands for a particular *number*, which can be calculated using the so-called "Feynman rules." I'm not going to explain how this is done – it's where the hard labor of particle physics resides. But the protocol is as follows. First you draw *all* the Feynman diagrams for the process in question, then you evaluate each diagram to get the associated number, and finally you add up the numbers to get the total result. Thus for electron–positron scattering we would draw the following diagrams:

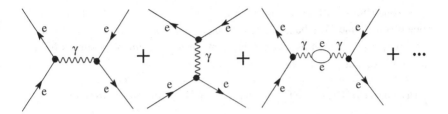

Oops! There are (pretty obviously) infinitely many diagrams! Luckily, in the evaluation of a diagram, each vertex throws in a factor of the **fine structure constant,**

$$\alpha = \frac{2\pi k e^2}{hc} = \frac{1}{137}, \tag{4.14}$$

where k is the constant in Coulomb's law (Eq. (1.18)), e is the charge of the electron (Eq. (4.1)), h is Planck's constant (Eq. (3.2)), and c is the speed of light (Eq. (1.34)). Because this number is very small, the "higher-order diagrams," with more and more vertices, contribute less and less, and they can be ignored, if all you require is a good approximation. In electrodynamics, diagrams with more than six vertices are almost never included in the calculations, and yet the precision achieved is by far the best in all of physics.

Everybody gets excited when they learn that a line pointing opposite to the left represents an antiparticle. Does that mean antiparticles really *are* particles propagating "backwards in time" (whatever that means)? I don't think so. It's just a rule for interpreting the diagram. But I can't resist telling you a fanciful story, if you promise not to take it too seriously. Feynman claimed that his advisor, John Archibald Wheeler, offered the following explanation for why all electrons are identical: there's only *one* of them! Consider the following diagram.

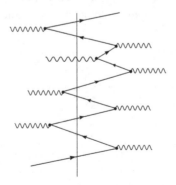

At a particular moment in time, represented by the vertical line, there are in existence a bunch of electrons (lines with arrows to the right) and a bunch of positrons (lines with arrows to the left) – and some photons. But there's only one electron line! Maybe all the electrons in the universe are connected up in this way. Of course, this means that there must be a lot of positrons hiding somewhere, but who knows... ?

Problem 27. Draw two Feynman diagrams for **Compton scattering**, $e^- + \gamma \rightarrow e^- + \gamma$.

Problem 28. Draw a Feynman diagram representing **pair annihilation**, $e^+ + e^- \rightarrow \gamma + \gamma$.

Problem 29. Draw the simplest Feynman diagram representing "Delbruck scattering": $\gamma + \gamma \rightarrow \gamma + \gamma$. (This process, the scattering of light by light, has no classical analog.)

Problem 30. Put in the actual values for k, e, h, and c, to calculate the fine structure constant (Eq. (4.14)). Show that it is a pure number (all the units cancel out).[33]

4.4.2 Chromodynamics

The strong forces are similar, except that the coupling is determined not by the electric charge of the particle (via the fine structure constant), but by its *color*. Because color plays a role analogous to that of charge in electrodynamics, the theory of the strong interactions is called **chromodynamics**. Leptons do not carry color, (they are, if you like, "color neutral"); accordingly, the strong force affects only the quarks.

The fundamental vertex is:

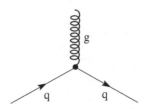

---

[33] Notice that α combines the basic constants from electrodynamics (k), special relativity (c), and quantum mechanics (h), together with fundamental unit of electric charge (e). It carries almost mystical significance to physicists, for whom an *ab initio* calculation of α stands as the ultimate holy grail.

Here a quark enters, emits or absorbs a gluon, and exits. Again, you snap these tinker toys together to represent all possible strong processes. For instance, one of the diagrams involved in the binding of a u and a $\bar{d}$ to make the π^+ is:

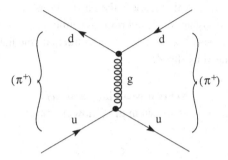

The decay of the Δ^0 baryon (ddd) into a proton (uud) and a π^- ($d\bar{u}$) is represented by the following diagram:

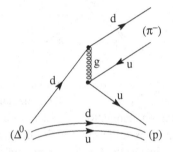

(Notice the two **spectator** quarks that go along for the ride, but do not actually participate in the interaction.)

Unfortunately, the parameter analogous to α in chromodynamics is *not* small, and that means that higher-order diagrams cannot in general be ignored. There are ways of working around this, in some cases, but the overall consequence is that calculations in chromodynamics are far more difficult, and much less reliable, than in electrodynamics.

Problem 31. Draw a Feynman diagram representing the decay of the Δ^- (ddd): $\Delta^- \to n + \pi^-$.

4.4.3 Weak interactions

Only *charged* particles experience electromagnetic forces; only *quarks* experience strong forces; but weak interactions are universal. They are the only forces that affect neutrinos, and (as the name suggests) they are far weaker than the other two, which is why neutrinos interact so feebly with matter. The weak interaction vertex factor is so small that one almost never has to consider higher-order diagrams; in this respect the calculations are a lot simpler than electrodynamics (to say nothing of chromodynamics).

There are two types of weak interaction: **charged** (mediated by the $W^{\pm}$), and **neutral**, mediated by the Z.

Neutral weak interactions

The fundamental process is shown in the following diagram:

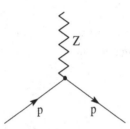

The particle p can be any quark or any lepton. Notice that every process mediated by the photon could just as well be mediated by the Z – there is, one might say, "weak contamination" in all electromagnetic interactions. But the weak contribution is ordinarily negligible. On the other hand, the Z couples also to neutrinos, allowing reactions such as neutrino–electron scattering.

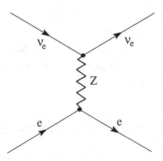

Charged weak interactions

The fundamental process is as

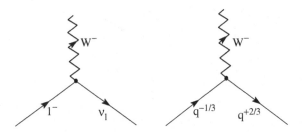

In the first diagram a charged lepton (e, μ, or τ) converts into the corresponding neutrino, with the emission of a W^-; in the second diagram a quark of charge $-1/3$ (d, s, or b) converts into a quark of charge $+2/3$ (u, c, or t), with the emission of a W^-. (Because the W^+ is the antiparticle of the W^-, it doesn't matter whether you draw the mediator as a W^- with the arrow pointing upward, or a W^+ with the arrow pointing down – it's the same thing.) The reverse process, $\nu \rightarrow l^- + W^+$ or $q^{2/3} \rightarrow q^{-1/3} + W^+$, is also allowed.

The charged weak interactions are the only ones that change the *type* ("flavor") of the quark or lepton (an electron turns into a neutrino, for example, or a d quark becomes a u quark). They are responsible for all true[34] particle decays. For example, here is the diagram representing the decay of the muon ($\mu \rightarrow e + \nu_\mu + \bar{\nu}_e$):

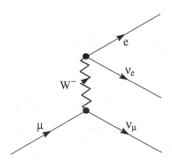

The decay of the neutron ($n \rightarrow p + e + \bar{\nu}_e$)[35] is strikingly similar, except that this time it is d going to u, instead of μ to ν_μ (and there are a couple of spectators):

---

[34] This is my personal view, and it is not shared by most physicists. It's purely a matter of terminology, but I prefer to say that a typical electromagnetic "decay," such as $\pi^0 \rightarrow \gamma + \gamma$, is really just the annihilation of a quark/antiquark pair; a strong "decay," such as $\Delta^0 \rightarrow p + \pi^-$, is really just the creation of a $u/\bar{u}$ pair, and a rearrangement of the quarks.

[35] Notice, by the way, that it is actually an *anti*-neutrino that comes out, not a neutrino, as was originally assumed.

136   *Elementary particles*

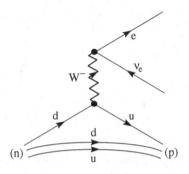

There is a curious difference between the charged weak interactions for quarks and for leptons: the charged lepton always converts to the neutrino *in the same generation* ($e \to \nu_e$, $\mu \to \nu_\mu$, and $\tau \to \nu_\tau$), but the quarks sometimes cross generations ($d \to c$, for example, or $s \to u$). This allows for weak processes such as the decay of the lambda ($\Lambda \to p + \pi^-$), in which an s goes to a u.

Problem 32. The π^- meson decays (by the weak interaction) to an electron and an antineutrino: $\pi^- \to e + \bar{\nu}_e$. Draw the Feynman diagram for this process.

Problem 33. Draw a Feynman diagram representing the process $\Lambda \to p + \pi^-$.

4.4.4 Conservation laws

According to the Standard Model, then, all physical processes[36] derive from three fundamental vertices:

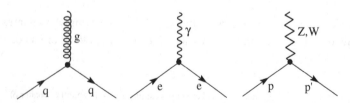

---

[36] Except gravity, of course. And there are some vertices I have not discussed, involving only the mediators (no quarks or leptons). For example, the W^+ and W^- are charged, so they couple to the photon ($W^\pm \to W^\pm + \gamma$), and there are even some four-mediator vertices, such as $W^+ + W^- \to Z + Z$.

In chromodynamics, a quark (q) emits/absorbs a gluon (g); in electrodynamics, a charged particle (e) emits/absorbs a photon (γ); and in the weak interactions, any quark or lepton (p) emits/absorbs a Z or a W, converting (in the case of the W) into a different quark or lepton (p'). Snap these together, in all possible combinations, and you generate all possible reactions. But there are some processes that *never* occur, because there is simply no way to produce them from the three allowed vertices. It is useful to codify these "illegal" reactions in the form of conservation laws.[37]

(1) **Energy and momentum.** Every process must conserve (relativistic) energy and momentum. A particle cannot, for example, decay into heavier particles (all you've got to work with is the rest energy of the original particle, and that's not enough to make the rest energies of the final particles).[38] You may be able to draw a Feynman diagram, but evaluation of that diagram would give zero. For example, $e \to \nu_e + W^-$ is represented by the simple diagram

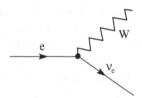

but it's not going to happen.

On the other hand, a *collision* between two particles *can* produce heavier products, if you give them enough kinetic energy to cover the increase in rest energy. For instance, in **pair production** ($\gamma + \gamma \to e^+ + e^-$) the incoming rest mass is *zero*, but it happens all the time, as long as the photon energies exceed a certain "threshold" value.

(2) **Electric charge.** All of the primitive vertices conserve electric charge (if charge goes in, an equal amount of charge comes out), and hence any diagram you generate by snapping them together will also conserve charge. The electron cannot decay, for example, because there is no lighter charged

---

[37] I am doing some violence to history here. These conservation laws were discovered empirically long before the interactions were understood, and some would say they are more fundamental than the vertices themselves. Certainly this is true of energy and momentum, and probably of charge.

[38] You might hope to beat the rap by accelerating the decaying particle up to high speed, and using its *kinetic* energy to make up the difference. It doesn't work: the first postulate of special relativity authorizes you to examine the process in any inertial system you like, and if it's impossible in the rest frame of the original particle, it must be impossible in any other reference frame.

particle for it to decay into. Note that pair annihilation ($e^+ + e^- \rightarrow \gamma + \gamma$) does not violate conservation of charge, since the total was zero before and after the event.

(3) **Lepton number.** In all three vertices, if a lepton goes in, a lepton comes out (in the case of the charged weak interactions it's a *different* lepton – a neutrino replacing an electron, for instance – but the *number* is unchanged). What about pair annihilation? Well, we assign to the *anti*-leptons a *negative* lepton number, so in this case the initial total is $1 - 1 = 0$. But the original picture of neutron decay, $n \rightarrow p + e$, is impossible in the Standard Model; the actual process, as we now know, is

$$n \rightarrow p + e + \bar{\nu}_e,$$

and it's got to be an *anti*-neutrino, to make the outgoing lepton number zero.

Because the charged weak interactions for leptons do not cross generations, we can actually say something stronger: lepton number is separately conserved for *each generation*. There's an **electron number** ($+1$ for e^- and ν_e, -1 for e^+ and $\bar{\nu}_e$, and zero for everything else), a **muon number** ($+1$ for μ^- and ν_μ, -1 for μ^+ and $\bar{\nu}_\mu$, and zero for everything else), and a **tau number** ($+1$ for τ^- and ν_τ, -1 for τ^+ and $\bar{\nu}_\tau$, and zero for everything else), and all three of them are conserved. Thus, for example, when the positive pion decays into a muon and a neutrino,

$$\pi^+ \rightarrow \mu^+ + \nu_\mu,$$

it's got to be a *muon* neutrino (not an electron neutrino), and when the muon decays into an electron and two neutrinos, one of them has to be a muon neutrino, and the other an electron antineutrino:

$$\mu \rightarrow e + \bar{\nu}_e + \nu_\mu.$$

In the early days it was a mystery why the muon doesn't decay to an electron plus a photon:

$$\mu \not\rightarrow e + \gamma,$$

and conservation of the separate lepton numbers was invented as an "explanation." (Here the muon number goes from 1 to 0, and the electron number from 0 to 1; the *total* lepton number is conserved, but not the individual electron and muon numbers.) Of course, this didn't really explain anything, but it did guide the development of the theory of weak interactions.

The recently discovered neutrino oscillations *do* cross generations. Electron neutrions convert into muon or tau neutrinos (and vice versa):

$\nu_e \leftrightarrow \nu_\mu \leftrightarrow \nu_\tau$. But this occurs "in flight," not at a weak vertex, and is significant only over very large distances – for neutrinos coming from the Sun, say, or from the upper atmosphere. It is sometimes claimed that neutrino oscillations represent physics "beyond the Standard Model." I disagree. Neutrino oscillations (and nonzero neutrino masses) fit very comfortably within the Standard Model. In fact, they make the theory more consistent, since leptons, like quarks, can now cross generations.

(4) **Baryon number.** Similarly, all vertices conserve **quark number** ($+1$ for a quark, -1 for an antiquark) – if a quark goes in, a quark comes out. (In the case of the charged weak interactions, it's not the *same* quark, but never mind – we're only talking about the total number.) Since free quarks don't occur in nature, it's more convenient to count *baryons*. **Baryon number** is just quark number divided by 3. Notice that *mesons* have quark number zero (one quark and one antiquark), and you can create any number of them, as long as you have enough energy:

$$p + p \rightarrow \begin{cases} p + p + \pi^0 \\ p + p + \pi^+ + \pi^- \\ p + p + \pi^0 + \pi^0 + \pi^0 \\ p + p + \pi^+ + \pi^- + \pi^0 + \pi^0, \end{cases}$$

etc. But conservation of baryon number forbids, for example, the decay of the proton:

$$p \not\rightarrow e^+ + \nu_e$$

(luckily for us).

(5) **Quark flavor.** In the case of the strong, electromagnetic, and neutral weak, interactions, the *same* particle (quark or lepton) comes out as went in[39] – accompanied now by a gluon or a photon or a Z. But the charged weak interactions change the quark flavor ($d \leftrightarrow u$, for example, or $s \leftrightarrow c$). If you know the process is strong or electromagnetic, then, you can be sure that flavor (strangeness, charm, beauty, truth – or for that matter "up-ness" and "down-ness," though these terms are never used) is conserved. But if it's a charged weak interaction, then flavor is *not* conserved.

Because the weak interactions are so feeble, we say that flavor is "approximately" conserved. Strangeness, for instance, is conserved in the production process (that's the observation that led Gell-Mann to introduce

---

[39] Well . . . I don't know if it's literally the same particle – but it is certainly the same *type* of particle.

the term), because this is a strong interaction.[40] But strangeness is *not* necessarily conserved in decays that involve the charged weak interactions, such as $\Lambda \rightarrow p + \pi^-$; in fact, the nonconservation of flavor is a flag, telling you that the reaction in question must be due to a charged weak interaction.[41]

Problem 34. Examine the following processes, and state for each one whether it is *possible* or *impossible*, according to the Standard Model. In the former case, state which interaction is responsible (strong, electromagnetic, or weak);[42] in the latter case cite a conservation law that forbids the process. (Following the usual custom, I will not always indicate the charge when it is unambiguous; thus γ (photon) and n (neutron) are neutral, e (electron) and μ (muon) are negative, p (proton) is positive, etc.)

(1) $p + \bar{p} \rightarrow \pi^+ + \pi^0$.

(2) $\mu^+ + \mu^- \rightarrow e^+ + e^-$.

(3) $p \rightarrow e^+ + \gamma$.

(4) $\pi^0 \rightarrow \gamma + \gamma$.

(5) $\mu \rightarrow e + \nu_\mu$.

(6) $p + p \rightarrow p + p + p + \bar{p}$.

Problem 36. Suppose two photons approach one another from opposite directions, with equal and opposite momenta, creating an electron/positron pair ($\gamma + \gamma \rightarrow e^- + e^+$). What is the energy threshold (that is, the minimum possible photon energy) for this process?

Problem 37. Members of baryon decuplet typically decay, via the strong interaction, to a member of the baryon octet plus a light meson (π or K). (a) List the likely decay modes for the Δ^{++}, the Δ^+, the Δ^0, and the Δ^-. (b) Do the same for the Σ^{*+} and the Ξ^{*0}.[43] (c) How about the Ω^-? When Gell-Mann was predicting the existence of the Ω^-, he knew that its

---

[40] Could you produce strange particles by the weak force? I suppose so, but it would be extremely rare, compared to the copious production by the strong force.

[41] In the laboratory, typical strong decays have lifetimes around 10^{-23} s, electromagnetic decays are around 10^{-16} s, and weak decays range from 10^{-13} s to 15 minutes (for the neutron).

[42] If there's a photon involved, it's got to be electromagnetic; if there's a neutrino involved, it's got to be weak; if flavor is not conserved, it must be charged weak.

[43] Don't confuse these with the Σ and the Ξ – those are octet particles. I'm afraid we're running out of Greek letters, so the star is there to distinguish them.

rest energy should be about 1670 MeV, while the rest energy of the Ξ is 1320 MeV, and that of the K is 495 MeV. He concluded that it could not decay, as the others do, by the strong force – it would have to decay by the weak interaction, and would therefore have a much longer lifetime than the other decuplet baryons. In point of fact, it decays to $\Lambda + K^-$ or $\Xi + \pi$, violating conservation of strangeness. Explain Gell-mann's reasoning.

4.4.5 Unification

For centuries, **unification** has been a recurring theme in physics. We seem, naively, to be surrounded by many different kinds of forces, but as our understanding and sophistication grow, we come to recognize more and more of them as different aspects of the same thing. Newton's inspiration that the force causing an apple to fall is the same as the force that holds planets in their orbits (gravity) was perhaps the first great unification, but in the nineteenth century electricity and magnetism were unified in a single theory, electromagnetism, and by the middle of the twentieth century *all* the forces of nature had boiled down to just four: strong, electromagnetic, weak, and gravitational.

Then, in the 1960s, Sheldon Glashow, Steven Weinberg, and Abdus Salam showed that the weak and electromagnetic interactions could be regarded as two parts of a unified **electro-weak** theory. The obvious objection, that the weak force is so much *weaker*, is due, it turns out, not to any intrinsic difference in strength, but to the fact that the weak mediators (the Ws and the Z) are so very heavy.

Indeed, the enormous mass of the W and the Z, when all other mediators are *massless*, was profoundly perplexing, and held up the development of the theory. The widely accepted resolution was the so-called **Higgs mechanism**, named for Peter Higgs (though others had essentially the same idea, in the same year – 1964). The Higgs mechanism requires the existence of a **Higgs boson**, a particle that couples also to the quarks and leptons, and is supposed to endow them all with their mass. Awkwardly, however, the Higgs particle has never been detected in the laboratory, and it is now nearly 50 years since it was proposed. Presumably it is just too heavy to have been produced by any existing accelerator, but if it is not found soon in the **Large Hadron Collider** at CERN – a machine built specifically for this purpose – there will be much consternation among particle physicists.[44]

---

[44] By the summer of 2012 there was strong (though still not absolutely conclusive) evidence for the Higgs, at a mass of 126 000 MeV/c^2.

The next step, obviously, would be to unify the strong and electro-weak forces in a **Grand Unified Theory**. Several candidate GUTs have been proposed. They suggest that the proton itself is unstable (though its predicted lifetime is much greater than the age of the universe).[45] The experiments appear to rule this out, so we have apparently not yet come up with the correct scheme. And, of course, the ultimate unification would include gravity, in one final **Theory of Everything**. Since 1984 the great hope has been **Superstring Theory**, with its bizarre pantheon of new particles (twins of the known particles, from "photinos" and "Winos" to "sneutrinos" and "squarks") and extra dimensions. But all this is in the nature of speculation, and way beyond the established facts of the Standard Model.

---

[45] Because of the statistical nature of the decay process, that doesn't mean that *no* proton would have decayed yet – some decay sooner, and some later, than the nominal lifetime.

5

Cosmology

Cosmology is the study of the structure and evolution of the universe as a whole. Prior to the twentieth century this was the domain of religion and speculative philosophy, but it is now a serious branch of physics, and one with spectacular implications. The transition from myth to science was brought about, in large measure, by two empirical findings: Hubble's observation (in 1929) that the Universe is expanding, and the discovery by Penzias and Wilson of Cosmic Microwave Background radiation (in 1965). Taken together, they indicate that the Universe as we know it began with a gigantic explosion 13.7 billion years ago – the **Big Bang**. In the next two sections we will explore these developments.

But first I would like to alert you to a fundamental assumption that informs almost all modern thinking about cosmology. Before Copernicus, most people assumed that the Earth is at the center of the Universe. The Sun, the Moon, the planets, and the stars orbit around us. Copernicus showed, to the contrary, that while the Moon orbits the Earth, the planets (including the Earth) orbit the *Sun*. And we now know that the Sun itself is just one of billions and billions of stars. The Earth is *not* at the center of the Universe – nor is the Sun. Modern cosmology takes the Copernican revolution to its logical conclusion: The Universe *has* no center, or any kind of "preferred" location – at a given time, it is the same in all places and in all directions.[1] This is known as the **Cosmological Principle**.

Of course, on a relatively small scale the Cosmological Principle is not true at all. In one direction I see Mars, in another direction I see the Moon, and in a third the sky is completely black. But the assumption is that if I average over

---

[1] In technical language, the Universe is **homogeneous** (the same everywhere) and **isotropic** (the same in all directions).

large enough regions, these local variations will smooth out, and the Universe will appear featureless and uniform – like an enormous pine forest viewed from space: up close there are clumps of trees here and there, with little meadows in between, but from above it just looks like a pale green carpet. We might speak of aggregate quantities, such as the number of trees per acre, or the number of owls per square mile, but the actual location of individual trees and owls would be quite irrelevant.

What are the contents of this "smoothed-out" universe? At this stage in cosmic evolution, the distribution seems to be as follows:

Type	Energy density (J/m^3)	Percent
Photons	5×10^{-14}	0.005%
Neutrinos	9×10^{-13}	0.1%
Atoms	4×10^{-11}	4.6%
Dark matter	2×10^{-10}	23%
Dark energy	6×10^{-10}	72%
Total	9×10^{-10}	100%

How did it come to be so, and what is all this "dark" stuff? That will be the subject for Sections 5.3 and 5.4.

Problem 1. Assuming all the atoms in the Universe are hydrogen (which is not far from the truth), how many atoms are there per cubic meter, on average? [Assume they are nonrelativistic, so their energy is almost entirely rest energy.]

5.1 Expansion of the Universe

5.1.1 Stars and galaxies

When you look up at the night sky, you see thousands of stars, each one more or less like the Sun, only farther away. But with a powerful telescope you would see much larger – though vastly more distant – objects that appear fuzzy; they are called **galaxies**:

Each galaxy contains billions of stars:

The stars we see with the naked eye belong to our own Galaxy, the **Milky Way**:

The Solar System resides toward the edge of the Milky Way, in one of the spiral arms:

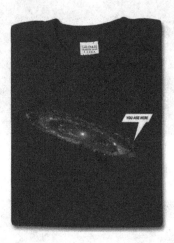

5.1.2 The cosmological redshift

Early in the twentieth century, it was noticed that the spectra of almost all galaxies are shifted toward the red (longer wavelength).

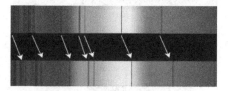

Each spectral line (such as those in the Balmer series for hydrogen) occurs at a larger wavelength, in the light from a distant galaxy, as compared with measurements in the laboratory.

This **cosmological redshift** is presumably due to the **Doppler effect**, which stretches out waves from a receding object:

eye   source at rest   eye   source receding

(The Doppler effect is what makes a police siren sound higher-pitched when it is coming toward you, and lower when it's going away – only in our case we're talking about light, rather than sound.) The fact that spectral lines are shifted toward the red suggests that these galaxies are *moving away from us*.[2]

But if everything is moving away from us, doesn't that mean we are, after all, at the "center"? Not necessarily. What if the Universe is like a giant chessboard, with a galaxy on every square? If the board is somehow expanding, then every galaxy is moving away from every other galaxy, and yet (if the board goes on forever) there is no "center." If you don't like the idea of an infinite Universe, picture it instead as the surface of a huge balloon, with the galaxies pasted on it like sequins. If the balloon is being inflated, all the sequins move away from all the others, but there is no "center" (at any rate, not on the surface). These two images are not supposed to be realistic models of the Universe, of course; they are intended only to disabuse you of the naive conclusion that if all galaxies are moving away from us, we must be at the center. That simply does not follow.

What the cosmological redshift *does* suggest is that the Universe is expanding. What exactly does this mean? Is it space itself that is expanding, or is it just that galaxies are flying apart *through* it, like the pieces of a firecracker that has just exploded? In the chessboard model it would be hard to say whether the galaxies are nailed to an expanding board or drifting apart over a static board, but in the balloon model it cannot be that they are moving over the surface of a static balloon, for in that case they wouldn't all be moving away – some of them, inevitably, would be *approaching*. If they are *all* moving apart, it must be that the balloon itself is expanding. Modern cosmology holds that it is space itself that is doing the expanding, and the galaxies are simply going along for the ride.

But if absolutely *everything* were expanding, how would we ever know it? If you woke up in the morning three times your normal height, but everyone else was taller too, and all your furniture, and your meter sticks – *everything* – how could you tell? That's not what's happening here. The galaxies themselves are *not* expanding – their size is dictated by the laws of physics.[3] It's the space between them that is expanding.

---

[2] If the spectra had been shifted toward the *blue* (shorter wavelengths), we would have concluded that the galaxies are coming *toward* us. In point of fact a very small number of nearby galaxies *are* slightly blue-shifted, but this is presumably just random motion, that tells us nothing about the expansion or contraction of the Universe as a whole.

[3] If you want a picture to keep in mind, think of the sequins pasted to the expanding balloon. Nobody's changing the size of the *sequins* – that was fixed by the manufacturer; what *is* expanding is the space they live in (the surface of the balloon).

5.1.3 Hubble's law

In 1929 Edwin Hubble[4] published what we now call **Hubble's law**, which says
that the speed (v) of a galaxy, as it recedes from us, is *proportional to how far
away it is* (d):

$$v = Hd. \tag{5.1}$$

The proportionality factor, H, is called **Hubble's constant**; its present value is
about 0.02 m/s per light-year.[5] "Constant" is a serious misnomer – at any given
moment it is the same for all distances d, but it certainly *does* change with time,
as we shall see.

Actually, Hubble's law is the *only* possible expansion rule consistent with the
Cosmological Principle. Imagine two galaxies (A and B, in the figure below) a
distance d away. They are receding from us at speed v. Picture also two other
galaxies (C and D) that are twice as far away; according to Hubble's law they
are receding at $2v$.

How do things look from the perspective of galaxy A? Subtracting v from all
the velocities,[6] I get the following.

Exactly the same – as required by the Cosmological Principle. But if Hubble's
law had said, for instance, that the velocity is proportional to the *square* of the
distance, then C and D would be receding from us at $4v$,

---

[4] Like most major scientific discoveries, the provenance of Hubble's law is not quite as
straightforward as we like to pretend. The formula itself was proposed (largely on theoretical
grounds) by Georges Lemaître in 1927, and much of the redshift data came from Vesto Slipher.
Hubble's main contribution was the determination of galactic distances, and he was hesitant to
interpret his results as proving the expansion of the Universe.

[5] A **light-year** (ly) is the distance light travels in one year: 9.46×10^{15} m. it's a convenient unit
for astronomical lengths.

[6] I'll use Galileo's velocity addition rule; the speed of most galaxies is much less than c.

$$\xleftarrow{-9v} \circ \xleftarrow{-4v} \circ \xleftarrow{-v} \circ \quad \circ \quad \circ \xrightarrow{v} \circ \xrightarrow{4v} \circ \xrightarrow{9v}$$

| F | D | B | (us) | A | C | E |

and therefore C would be going at $3v$ with respect to A.

$$\xleftarrow{-5v} \circ \xleftarrow{-2v} \circ \xleftarrow{-v} \circ \quad \circ \quad \circ \xrightarrow{3v} \circ \xrightarrow{8v} \circ \xrightarrow{15v}$$

| D | B | (us) | A | C | E |

The Universe would look very different from the perspective of galaxy A, in violation of the Cosmological Principle.

Although measuring velocity is easy and accurate (using the redshift of spectral lines), determining the distance to galaxies is notoriously difficult and unreliable. (As a result, estimates of Hubble's constant tend to jump around as the technology improves.) Why are astronomical distances so difficult to measure? After all, an object that is farther away appears dimmer: apparent brightness is inversely proportional to the square of the distance. If you had two 100 watt bulbs, and one of them was twice as far away, it would appear a quarter as bright. This should be *easy!* The trouble is, of course, that we don't know the wattage of distant galaxies. How can you tell whether it's a very bright object that is far away, or a dim object that is quite close? What we need is a **standard candle** – a type of star whose intrinsic brightness is always the same. Then from the *observed* brightness we would immediately know how far away it is – and the distance to the galaxy in which it resides.

Luckily, there exist such standard candles. Two, in particular, have been instrumental in the development of modern astronomy: **Cepheid variables** (stars whose brightness varies periodically, at a frequency that is correlated to their intrinsic brightness), and **Type 1a supernovae** (exploding **white dwarf** stars). The details do not concern us, but these "candles" are far from perfect, and I think it is fair to say that reliable distance measurements remain the Achilles' heel of modern cosmology.

Problem 2. (a) Calculate the length of a light year, in meters. (b) Astronomical distances are sometimes given in **parsecs** (pc); a parsec is 3.262 light-years (ly). How many meters are there in a parsec?

Problem 3. The Virgo galaxy is 50 million light years away; what is its speed, relative to Earth?

Problem 4. Redshift measurements indicate that a certain galaxy is moving away from us at half the speed of light. How far away is this galaxy, in light-years?

Problem 5. Here is a graph of actual redshift data, with the velocity (v) of various galaxies (in km/s) plotted against their distance (d), in Mpc = 10^6 pc (see Problem 2). What feature of this graph supports Hubble's law? From the graph, determine Hubble's constant.

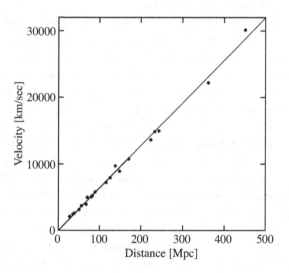

5.1.4 The Big Bang

Hubble's law is exactly what you would expect if the Universe started out with a gigantic explosion, like a firecracker. The bits and pieces would fly apart with various speeds, and after a time t each one would have traveled a distance $d = vt$, which is to say, $v = (1/t)d$. This is Hubble's law, with $H = 1/t$. Evidently Hubble's constant is the *reciprocal of the time since the explosion.* We call this cosmic explosion the **Big Bang**, and the time since the Big Bang is the **age of the Universe**:

$$t = \frac{1}{H} = \frac{\text{ly}}{0.02\,\text{m/s}} = \frac{(3 \times 10^8\,\text{m/s})\,\text{yr}}{0.02\,\text{m/s}} = 1.5 \times 10^{10}\,\text{years.} \tag{5.2}$$

That assumes the velocity of each galaxy has remained the same, ever since the Big Bang. Until recently it was universally assumed that the expansion actually slows down over time, because of the gravitational attraction of all matter. In that case the true "age of the Universe" would actually be somewhat *less* than 15 billion years. In any event, Hubble's "constant" *decreases* as the Universe grows older.

But this naive picture, with shrapnel flying out from one point through a preexisting static space, is inconsistent with the the Cosmological Principle. It would endow the Universe with an obvious "center": the location of that primordial firecracker. You must think, rather, of space itself expanding, carrying the galaxies along with it – a stretching infinite chessboard, say, or an inflating balloon. The Big Bang occurred everywhere, all at once. At the time of the Big Bang the distance between any two galaxies was zero (!), and the "age of the Universe" $(1/H)$ is the time that has elapsed since that magic moment.

5.2 The Cosmic Microwave Background

Hubble's discovery practically *begs* us to conclude that our Universe began with a giant explosion. And yet this seemed so outlandish that very few people took it seriously until 1965, when Penzias and Wilson – quite by accident – discovered the **Cosmic Microwave Background** radiation – in effect, the flash of light left over from the Big Bang. To explain their discovery and its implications, I must first tell you a bit more about blackbody radiation.

5.2.1 Blackbody radiation

On the **Celsius** temperature scale, ice melts at $0\,°C$, water boils at $100\,°C$, and **absolute zero** (the lowest possible temperature, at which, classically, all random thermal motion ceases) is $-273\,°$C. But it is more sensible to measure temperatures up from absolute zero – in effect, to add 273 to all Celsius temperatures. This gives us the **Kelvin** temperature scale, on which absolute zero is $0\,K$, ice melts at $273\,K$, and water boils at $373\,K$. From now on, I'll always use the Kelvin scale, unless otherwise indicated.

Every object radiates – gives off electromagnetic waves ("light," though it need not be in the visible range). This is called **blackbody radiation**. It's a real misnomer – the Sun is, in this sense, a "blackbody"! It should be called "thermal radiation," because it is due to the random thermal motion of the charged particles (especially electrons) of which the object is composed.[7]

---

[7] I'm not talking about *reflected* light. Most of the objects around us are reflecting some of the light that hits them (and absorbing the rest) – a blue shirt absorbs red and green and reflects blue, for example. That's an entirely different mechanism. To avoid confusion with reflected light, we might consider an object that absorbs *everything* – a perfectly black body (hence the name). But even objects that are *not* black give off thermal radiation.

Naturally, the hotter an object is, the more thermal radiation it emits. In fact, the *power* (energy per unit time) radiated by an object at (Kelvin) temperature T is given by the **Stefan–Boltzmann law**:

$$P = \sigma A T^4, \qquad (5.3)$$

where A is the surface area of the object, and σ is a constant,

$$\sigma = 5.67 \times 10^{-8}\ \text{watts/m}^2\ \text{K}^4.$$

Example 1. The temperature of the surface of the Sun is $T_s = 5800\ \text{K}$; its radius is $R_s = 6.96 \times 10^8\ \text{m}$. What, then, is the total power radiated by the Sun? Determine the temperature of the Earth.

Solution: The surface area of a sphere is $4\pi R^2$, so

$$P_s = \sigma(4\pi R_s^2)T_s^4.$$

It is this solar radiation that keeps us warm – but of course most of it misses the Earth. The fraction absorbed by the Earth is the area subtended by the Earth on a spherical surface at the radius of the Earth's orbit, divided by the total area of that sphere:

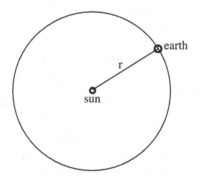

So the solar power absorbed by the Earth is

$$P_a = P_s \frac{\pi R_e^2}{4\pi r^2} = \sigma(4\pi R_s^2)T_s^4 \frac{\pi R_e^2}{4\pi r^2},$$

where R_e is the radius of the Earth, and r is the distance to the Sun.

Well, if the Earth is absorbing all this radiation, how come it doesn't heat up? *Answer:* The Earth *itself* radiates, by an amount

$$P_e = \sigma(4\pi R_e^2)T_e^4,$$

where T_e is the temperature of the Earth. In fact, since the Earth is *not* heating up (at least, not until recently), it must re-radiate exactly[8] the same amount of energy as it absorbs: $P_e = P_a$, or

$$\sigma(4\pi R_e^2)T_e^4 = \sigma(4\pi R_s^2)T_s^4 \frac{\pi R_e^2}{4\pi r^2} \quad \Rightarrow \quad T_e^4 = T_s^4 \frac{R_s^2}{4r^2},$$

so

$$T_e = T_s \sqrt{\frac{R_s}{2r}}. \tag{5.4}$$

This means the temperature of the Earth is determined by the temperature of the Sun, the radius of the Sun, and the distance to the Sun ($r = 1.50 \times 10^{11}$ m). Putting in the numbers:

$$T_e = 5800\sqrt{\frac{6.96 \times 10^8}{3.00 \times 10^{11}}} = 279 \text{ K},$$

or 6 °C (43 °F), which is roughly right (the measured average temperature of the Earth is about 14 °C).[9]

If every object gives off thermal (blackbody) radiation, how come people don't glow in the dark? Actually, they *do* – it's just that they don't glow in the visible range. A hotter object not only radiates *more*, it also radiates at *shorter wavelengths* (a hot poker radiates in the red, but a *very* hot object radiates more toward the blue – what we call "white hot"). The **spectrum** of blackbody radiation is given by the formula Planck discovered 1900; I won't write it out, but here is the graph, for three different temperatures:

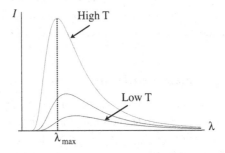

These curves are universal. At a given temperature, the radiation emitted is the same, regardless of what the object is made of. That's why, if you look into a red-hot kiln, you can hardly distinguish one thing from another – everything is giving off the same red glow. (Reflected light is totally different, and depends entirely on the chemistry of the surface.) In particular, the wavelength at which

---

[8] Some solar radiation is reflected, so P_a here is an overestimate.
[9] Equation (5.4) works pretty well for all the planets except Venus, which has a runaway greenhouse effect, and is considerably hotter than it "should" be.

the *most* power is radiated is given by **Wien's law**:

$$\lambda_{max} = \frac{2.90 \times 10^{-3}}{T} \text{ K m}. \tag{5.5}$$

The higher the temperature, the shorter the wavelength.

The Sun radiates predominantly in the visible[10] range; the Earth, being much cooler, radiates in the infrared (and so do people – that's why you can't see them in the dark, unless you use infrared goggles). Global warming presumably occurs because we have polluted the atmosphere with gases that are transparent in the visible (allowing the solar radiation in) but opaque in the infrared (blocking the Earth's radiation from getting out); because of this imbalance, the Earth really *is* heating up (that's the infamous **greenhouse effect**).

Problem 6. Room temperature is about 20 °C. What is this on the Kelvin scale?

Problem 7. (a) What is the total power radiated by the Sun (in watts)? (b) At what wavelength is the most power radiated? What color is this? (c) What is the total power radiated by the Earth? (d) At what wavelength does the Earth radiate most? What kind of radiation is this?

Problem 8. Human body temperature is 37 °C (=98.6 °F). Estimate the surface area of a person (this doesn't have to be very accurate), in square meters, and from this determine the total power radiated. At what wavelength does a person radiate most? What sort of radiation is this?

5.2.2 Penzias and Wilson

What Penzias and Wilson discovered is that we are bathed in radiation with a perfect blackbody spectrum corresponding to a temperature of 2.725 K. This is in the microwave region (λ_{max} is about a millimeter). It comes from all directions, and it does not fluctuate with the seasons or with day and night, so it is not coming from the Sun, or the Solar System, or even the Galaxy – and certainly not from any terrestrial source. Evidently its origin is the cosmos as a whole: it is thermal radiation left over from the Big Bang.

Not quite the Big Bang itself, but some 380 000 years *after* the Big Bang. In the early stages, the Universe was very dense and extremely hot – so hot that neutral atoms could not hold together; they existed only in ionized form, with

---

[10] Of course, the causal connection is the other way around: our eyes evolved to take maximum advantage of the radiation coming from the Sun.

free electrons roaming around at will, as they do in a metal. Such a Universe was completely opaque. But as it expanded, the Universe cooled, and when (after 380,000 years) it reached about 3000 K, neutral atoms began to form; the "fog" lifted, and electromagnetic radiation was liberated. What we are seeing today is the radiation left over from that era – except that, because of the continued expansion of the Universe, its wavelength has been red-shifted down from an effective temperature of 3000 K to slightly below 3 K.

Several visionaries (most notably George Gamow, in 1948) had predicted the existence of such remnant radiation, and even estimated its temperature, to surprising accuracy. But most serious physicists regarded this as science fiction, and Penzias and Wilson were not looking for anything of the sort – on the contrary, when they first discovered it they were irritated, thinking it was due to bird droppings on their antenna. But once it was correctly interpreted, their discovery of the Cosmic Microwave Background convinced all but the most tenacious skeptics that the Big Bang scenario is correct.

Problem 9. What Celsius temperature corresponds to the 2.725 K of the cosmic background radiation?

Problem 10. From the graph below (which is based on actual measurements of the cosmic background radiation), determine, as accurately as you can, the wavelength at which the intensity is greatest. Use this result to calculate the blackbody temperature of the radiation.

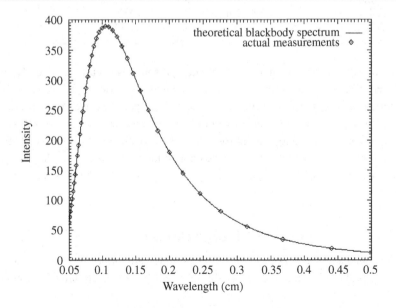

5.3 The origins of matter

The explosion itself began as pure energy, in the form of photons and other mediators. From them, by the mechanism of pair production (the reverse of pair annihilation), all the elementary particles (quarks and leptons) were produced, together with their antiparticles. This primordial soup was fantastically hot and incredibly dense. At first no bound states could form – no baryons or mesons, let alone nuclei or atoms – they would be blasted apart immediately by collisions. As the Universe cooled, quarks were able to coalesce into protons and neutrons. Later on, the protons and neutrons combined to make light nuclei. Still later, these nuclei were joined by electrons to make atoms. Eventually, stars and galaxies condensed out. And finally, in the interiors of stars, and in the spectacular explosion of dying stars (supernovae), the heavy elements were made. One of the triumphs of Big Bang cosmology is that we now have a pretty good idea of how each stage in this remarkable saga played out.

Object formed	Time after Big Bang	Temperature of Universe
Mediators	10^{-40} seconds	10^{30} K
Quarks and leptons	10^{-35} seconds	10^{26} K
Baryons and mesons	10^{-6} seconds	10^{11} K
Nuclei	3 minutes	10^9 K
Atoms	380 000 years	3,000 K
Stars	150 million years	100 K
Galaxies	500 million years	40 K
Solar System	8 billion years	10 K
People	13.7 billion years	2.73 K

The dominance of matter over antimatter presumably occurred very early, in the era of quark and lepton formation. Known asymmetries in the weak interactions can account for a slight preponderance of particles over antiparticles, and subsequent pair annihilation would leave a Universe of matter. But this is nowhere near enough to account for the fact that so much matter survived (calculations indicate that there should be enough matter to make a single galaxy, at most). So some other mechanism must be involved, but what it is we do not know.

5.3.1 Light elements

The light nuclei (deuterium, helium, and lithium) were formed from protons and neutrons about 3 minutes after the Big Bang, when the temperature was around a billion K. The process was nuclear **fusion**; some of the relevant reactions are:

$$n + p \to {}^2\mathrm{H} + \gamma$$
$$^2\mathrm{H} + p \to {}^3\mathrm{He} + \gamma$$
$$^3\mathrm{He} + {}^3\mathrm{He} \to {}^4\mathrm{He} + p + p$$
$$^3\mathrm{H} + {}^4\mathrm{He} \to {}^7\mathrm{Li} + \gamma.$$

The observed abundances of hydrogen, deuterium, and the isotopes of helium tell us a lot about conditions in the immediate aftermath of the Big Bang. Ordinary hydrogen is by far the most common element in the Universe, followed by helium (about 25% by mass); deuterium is about 0.002%.

5.3.2 Heavy elements

The heavier elements were not produced in the Big Bang, but by a complicated chain of fusion reactions occurring in the interiors of stars. When dying stars collapse and explode as supernovae, these elements are released into the surrounding space and subsequently incorporated into the next generation of stars and planets. It is a curious fact that you and I are made from the ashes of old stars, and (in the case of hydrogen) from remnants of the Big Bang.

5.3.3 Stars and galaxies

If the Cosmological Principle were perfectly correct (at all scales), there would be no stars or galaxies – just a uniform featureless smear of matter. But suppose somehow one region was slightly denser than its surroundings; then its extra gravitational attraction would suck in nearby matter, and it's not hard to imagine that this would lead to the clumping of material into larger structures. As in the formation of raindrops, which condense around specks of dust, all it takes is a "seed." But how did these seeds get started? Studies of the Cosmic Microwave Background show that it is extraordinarily uniform – the temperature in different directions is the same to better than one part in a thousand. In the figure below

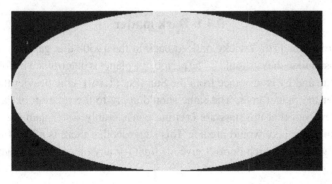

the temperature of the background radiation is indicated on a grey scale. It is amazingly uniform. (The oval part represents the entire celestial sphere – the far right wraps around and joins onto the far left, and the top attaches to the bottom.)

And yet, as indicated in the next figure (which exaggerates any deviations from the average), there are minute fluctuations, dating back to the time that atoms formed and the Universe became transparent. What caused them is not entirely clear, but they do appear to be sufficient to account for the eventual aggregation of stars and galaxies as we know them today.

5.4 Remaining mysteries

So far, the story is all very tidy. But there are some loose ends. It is now pretty clear that the matter we *know* about – atoms and radiation, galaxies and stars – accounts for only a small fraction of the total. The rest is "dark matter" and "dark energy." Dark matter is concentrated in galaxies, as a kind of spherical "halo"; dark energy is distributed uniformly throughout the Universe. What is the nature of this invisible stuff? There are some intriguing speculations, but very little hard evidence. And what is the *shape* of the Universe? Is it finite, or infinite? We know a lot about the past (since the Big Bang), but what does the future hold? These are all questions that may well be answered in the coming decade.

5.4.1 Dark matter

The astronomer Fritz Zwicky noticed back in the 1930s that galaxies tend to rotate faster than they "should." The period of a planet is determined by the mass of the Sun and by its distance from the Sun (Eq. (1.16)) – the heavier the Sun, the faster the planet goes. The same should apply to the rotation of galaxies, but it turns out that the stars are orbiting considerably faster than the *visible* matter in the galaxy would predict. This suggests that there is a lot of *unseen* mass in galaxies, which doesn't give off light (or any other form of radiation, as far as we can tell). And it is not a small amount – there seems to be about five times as much of this **dark matter** as there is of the ordinary stuff. But at present we have very little idea what it might be. It exerts a gravitational pull on ordinary matter (that's how we know it's there), but apart from that it doesn't seem to interact much. Is it made of some new kind of Weakly Interacting (like neutrinos, but) Massive Particles (WIMPs), or perhaps MAssive Compact Halo Objects (MACHOs) – cold dead stars, for instance, that have burned up all their fuel and no longer radiate? Nobody knows.

But there is compelling independent evidence that dark matter really does exist. When light from a distant object passes by a galaxy, it is bent by the galaxy's gravitational field, just as it is when it passes through a glass lens. This **gravitational lensing** produces distorted images of the distant object, such as the circular streaks in this photograph:

The shape and position of these images can be used to determine the mass of the galaxy, which turns out (once again) to be substantially greater than the mass of the observable matter. Today, gravitational lensing is widely used to "map out" the dark matter in the Universe.

5.4.2 The shape of the Universe

Back in Section 1.3.1 we calculated the acceleration of gravity, g. In that story, mass (m) plays two distinct roles: it is a measure of inertia (in Newton's second law, $F = ma$), and it tells you how strongly two objects attract (in the law of universal gravitation, $F = GMm/R^2$). The two ms cancel out, and as a result all objects, regardless of their mass, fall at the same rate – a remarkable fact first noted by Galileo. The equality of **inertial mass** and **gravitational mass** is now known as the **equivalence principle**. In Newton's theory, it is just an extraordinary coincidence. Einstein didn't believe in coincidences, and his theory of gravity (known, misleadingly, as the **General Theory of Relativity**) was designed to account for the principle of equivalence. In Einstein's approach, gravity is not a force at all, but rather a manifestation of the *curvature* of spacetime.

Imagine a stretched rubber membrane; place a billiard ball at the center, pushing the membrane down:

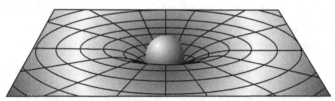

Now put a marble on the sheet – it will of course roll toward the center, accelerating as it goes. Or, if you give it some tangential speed, it will "orbit" around the billiard ball. From above, it looks as though the marble is *attracted* to the billiard ball, and we might solemnly announce a "law of attraction" (like Newton's) to describe the "force" pulling it in. But really, it's accelerated toward the center not because something there is pulling on it but because the surface on which it is moving slopes down that way.

This is a crude analog to a sophisticated four-dimensional theory, but it helps to convey the essence of Einstein's idea: large masses, such as the Earth, distort the space and time in their vicinity, and nearby objects, moving through that distorted region, accelerate – not because of some force acting on them, but because spacetime itself is curved. They try their best to obey Newton's first law (moving in a straight line at constant speed), but in a curved space there *are* no straight lines, and they travel instead on **geodesics**. (A geodesic is the path of shortest distance between two points.) These trajectories depend on the shape of the spacetime, but not on the mass of the moving particle – hence the equivalence principle.

Applying Einstein's theory to the Universe as a whole, and invoking the Cosmological Principle (at any given time the Universe in the large is the same at all places and in all directions), Friedmann found that that there are essentially three possible geometries: spherical, hyperbolic, and flat.

Which shape describes our own Universe depends on the density of matter. If the average density of the Universe is greater than a certain critical value, then the Universe closes on itself – it is finite and "spherical" (though this is really a spherical surface in four dimensions; it is hard to picture and impossible to render on a two-dimensional page). If the average density is less than the critical value, then the Universe is open ("hyperbolic") and infinite. And if the average density is right *at* the critical value, then the universe is flat (and infinite). The ratio of the actual density to the critical density is denoted by Ω_0 (Greek omega).

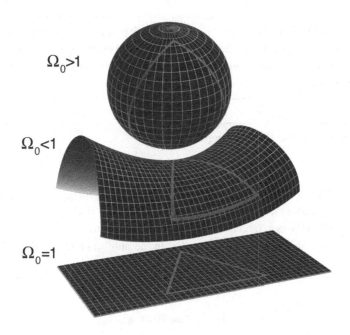

$\Omega_0 > 1$

$\Omega_0 < 1$

$\Omega_0 = 1$

The figure above shows two-dimensional surfaces that are spherical, hyperbolic, and flat. The gray triangles are constructed from geodesics. The sum of the angles of a triangle is greater than 180° in a spherical space, less than 180° in a hyperbolic space, and – of course – *equal* to 180° in a flat space.

The critical density is not very large – about 6 hydrogen atoms per cubic meter – but the *observed* average density (stars and galaxies, interstellar dust, radiation, and so on) is only a few percent of the critical value. Even if you include dark matter, it is nowhere near enough to "close" the Universe. Until recently, therefore, it seemed that the Universe must be open, and hyperbolic in shape. But there was always the possibility of "missing mass" out there that, for some reason, we could not detect, and some theories favor a flat Universe. The current evidence strongly suggests that the Universe is, in fact, *right at* the critical value, and hence is flat, and infinite.

Problem 11. The formula for the critical density is

$$\rho_c = \frac{3}{8\pi} \frac{H^2}{G}. \tag{5.6}$$

Put in the numbers, and calculate ρ_c in kg/m^3. How many hydrogen atoms per cubic meter does this amount to?

Problem 12. What is the critical *energy* density, in joules per cubic meter? (Use $E = mc^2$.) Compare the observed density of ordinary matter, from the table on page 143. What about the *total* density (including dark matter and dark energy)?

Problem 13. The **Hubble length**,

$$R = \frac{c}{H}, \tag{5.7}$$

is (roughly) the distance light has traveled since the Big Bang. It is sometimes called the **radius of the Universe**, though it's more like the radius of the *observable* Universe (and even that interpretation is naive); in any event, it sets the size scale for the Universe. Put in the numbers and calculate R, in meters, and in light-years.

5.4.3 The future

Until recently, it was generally assumed that the expansion of the Universe slows down over time, because of the gravitational attraction of all matter. This raises the intriguing question of whether the Universe will continue to expand (at a progressively slower rate), or whether it will eventually turn around and collapse back to a "big crunch." It depends (again) on how much matter there is out there. If the density exceeds the critical value (the same number, ρ_c, that determines whether it is open or closed), then it will collapse back; otherwise it will continue to expand forever. At any rate, that's what everyone believed until about 1998, when observations showed that the expansion is not slowing down at all, but *speeding up!* What can this possibly mean?

The story begins (as so many things in twentieth-century physics do) with Einstein. Soon after the publication of his General Theory of Relativity in 1916, he applied it to the cosmos as a whole, and was dismayed to find that it predicted an unstable (expanding or collapsing) Universe. This was well before Hubble, and Einstein thought it was absurd.[11] So he doctored up his theory with a "fudge factor," the notorious **cosmological constant** (Λ), which in effect introduced a long-range repulsive force to counteract the gravitational attraction and thus stabilize the Universe. When Hubble and the others discovered that the Universe is in fact expanding, Einstein repudiated the cosmological constant, calling it his "biggest blunder." If he had taken his own theory seriously, he could have *predicted* the expansion of the Universe a decade before Hubble.

But when evidence that the expansion is accelerating began to emerge, Einstein's fudge factor was dusted off and put back to work as an "explanation" for the large-scale repulsion.

5.4.4 Dark energy

Einstein conceived of Λ as a fundamental constant of nature, like Planck's constant or the speed of light – something that just *is* what it is, requiring no derivation or justification. In its modern reincarnation, physicists are more inclined to regard it as a dynamical quantity – an outward pressure counteracting the inward pull of gravity, attributable to some identifiable physical agency. Elementary particle theorists quickly convinced themselves that we should have expected a cosmological constant all along. Awkwardly, their calculations yield a value that is off by a factor of 10^{120}, which is surely the all-time record for a bad match between theory and experiment. In any event, the "stuff" responsible for the accelerated expansion has come to be called **dark energy**. It is supposed to permeate space uniformly (unlike dark matter, which is concentrated in galactic halos).

In the figure below, the bottom curve represents a closed, high-density Universe which expands for several billion years, then turns around and collapses under its own weight. The next curve represents a flat, critical-density Universe in which the expansion rate gradually slows down. The third curve shows an open, low-density Universe whose expansion is also slowing down, but not as much as the previous two because the pull of gravity is not as strong. The top curve shows a Universe with dark energy, which causes the expansion to speed up (accelerate). Ω_m is the density of matter (including dark matter), as a fraction of the critical density; Ω_v is the density of dark energy, as a fraction of the critical density. The observed accelerating expansion of the Universe suggests that we are on the uppermost curve.

---

[11] *Why* he thought it was so absurd is a mystery to me. Newton's theory of gravity also predicts an unstable Universe, for exactly the same reason: all matter attracts. But in 1917, everybody took it for granted that the Universe as a whole is static.

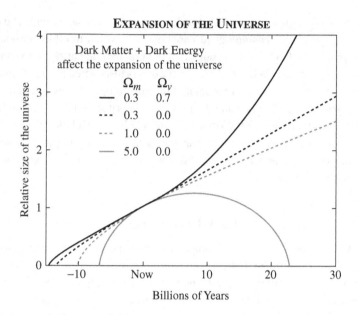

EXPANSION OF THE UNIVERSE

Dark Matter + Dark Energy affect the expansion of the universe

If all this is true, then matter as we know it accounts for a paltry 4.6% of the energy in the Universe, dark matter is another 23%, and the overwhelming majority (72%) is dark energy. (In the early Universe the proportions were quite

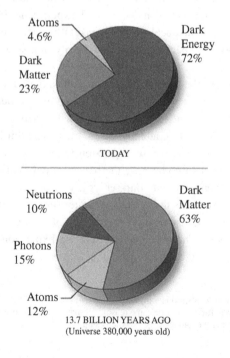

Atoms
4.6%

Dark Energy
72%

Dark Matter
23%

TODAY

Neutrions
10%

Dark Matter
63%

Photons
15%

Atoms
12%

13.7 BILLION YEARS AGO
(Universe 380,000 years old)

different.) Taken together, they are just about right to make the critical density, suggesting that the universe is flat, infinite, and destined to keep expanding forever. But the fundamental *nature* of dark energy – and dark matter – remains, at this point, a mystery.

Problem 14. If the Universe is really full of "dark energy," how come we don't notice it in everyday life? What is the mass equivalent of all the dark energy in the Earth? (Refer to the table on page 144 for the density of dark energy.)

Index

168 *Index*

Figure References

p. 32   Tables 1.1 and 1.2 reprinted by permission from D. J. Griffiths,
         Introduction to Electrodynamics, 3rd ed., Pearson, Boston (1999).
p. 35   Figure reprinted by permission from software program "Ripple Tank,"
         www.ottisoft.com.
p. 79   Diagram reprinted by permission from D. J. Griffiths, *Introduction to
         Quantum Mechanics, 2nd ed.*, Pearson, Boston (2005).
p. 87   Figure printed by permission using "Atom in a Box," v1.0.8, by Dauger
         Research, http://dauger.com.
p. 95   Figure reprinted from the Wikipedia article "Schrödinger's cat."
p. 108   Figure reprinted from the Wikipedia article "Binding energy curve."
p. 114   Reprinted from *The Study of Elementary Particles by the Photographic
         Method*, C. F. Powell, P. H. Fowler, and D. H. Perkins, Pergammon, New
         York (1959); first published in *Nature* **163**, 82 (1949).
p. 144(a) The Hercules Cluster. Reprinted with the permission of William Keel
           and Victor Andersen, who created the image, from
           http://apod.nasa.gov/apod/ap980827.html.
p. 144(b) The Andromeda Galaxy. Photo by Robert Gendler; reprinted by
           permission, from http://www.robgendlerastropics.com/M31Page.html.
p. 144(c) The Milky Way. By Steve Jurvetson (Flickr) [[CC-BY-2.0
           (http://creativecommons.org/licenses/by/2.0)], via Wikimedia Commons.
p. 145(a) ThinkGeek.com.
p. 145(b) Reprinted by permission from Harold Stokes,
           http://stokes.byu.edu/teaching_resources/computer_resources.html.
p. 149   Data from A. G. Reiss, W. H. Press, and R. P. Kirshner, *Astrophysical
         Journal* **473**, 88 (1996). Plotted by Ned Wright, and reprinted with
         permission from www.astro.ucla.edu/~wright/cosmo_01.htm.
p. 156   http://map.gsfc.nasa.gov/media/ContentMedia/990004b.jpg.
p. 157   http://map.gsfc.nasa.gov/media/101080/index.html.
p. 158   hubblesite.org/newscenter/archive/releases/2000/07/image/b/.
p. 160   http://map.gsfc.nasa.gov/universe/bb_concepts.html. (Also on front
         cover.)
p. 162   http://map.gsfc.nasa.gov/universe/bb_concepts.html.
p. 163   http://map.gsfc.nasa.gov/media/080998/index.html.

Fundamental constants

Newton's constant:	G	$= 6.6743 \times 10^{-11} \text{ m}^3/\text{kg s}^2$
Coulomb's constant:	k	$= 8.9876 \times 10^9 \text{ kg m}^3/\text{C}^2 \text{ s}^2$
Speed of sound:	v	$= 340 \text{ m/s}$
Hubble's constant:	H	$= 0.022 \text{ (m/s)/ly}$
Age of Universe:	$1/H$	$= 1.4 \times 10^{10} \text{ yr}$
Planck's constant:	h	$= 6.626 \times 10^{-34} \text{ J s}$
Speed of light:	c	$= 2.9979 \times 10^8 \text{ m/s}$
Mass of the electron:	m_e	$= 9.1094 \times 10^{-31} \text{ kg}$
Mass of the proton:	m_p	$= 1.6726 \times 10^{-27} \text{ kg} = 1.0073 \text{ u}$
Mass of the neutron:	m_n	$= 1.6749 \times 10^{-27} \text{ kg} = 1.0087 \text{ u}$
Charge of the proton:	e	$= 1.6022 \times 10^{-19} \text{ C}$
Charge of the electron:	$-e$	$= -1.6022 \times 10^{-19} \text{ C}$
Fine structure constant:	α	$= 2\pi k e^2/hc = 1/137.036$
Bohr radius:	a	$= h^2/4\pi^2 m_e e^2 = 5.2918 \times 10^{-11} \text{ m}$

Conversion factors

1 in	$= 2.54 \text{ cm}$
1 ly	$= 9.4605 \times 10^{15} \text{ m}$
1 pc	$= 3.0857 \times 10^{16} \text{ m}$
1 yr	$= 3.156 \times 10^7 \text{ s}$
1 u	$= 1.6605 \times 10^{-27} \text{ kg}$
1 MeV/c^2	$= 1.7827 \times 10^{-30} \text{ kg}$
1 eV	$= 1.602 \times 10^{-19} \text{ J}$
1 kwh	$= 3.600 \times 10^6 \text{ J}$

Greek alphabet

alpha	α	iota	ι	rho	ρ
beta	β	kappa	κ	sigma	σ, Σ
gamma	γ	lambda	λ, Λ	tau	τ
delta	δ, Δ	mu	μ	upsilon	υ, Υ
epsilon	ϵ	nu	ν	phi	ϕ
zeta	ζ	xi	ξ, Ξ	chi	χ
eta	η	omicron	o	psi	ψ
theta	θ	pi	π	omega	ω, Ω

Metric system prefixes

giga	G	billion	10^9
mega	M	million	10^6
kilo	k	thousand	10^3
centi	c	hundredth	10^{-2}
milli	m	thousandth	10^{-3}
micro	μ	millionth	10^{-6}
nano	n	billionth	10^{-9}

Astronomical data

Mass of the Earth :	5.97×10^{24} kg
Mass of the Moon :	7.36×10^{22} kg
Mass of the Sun :	1.99×10^{30} kg
Distance to the Moon :	3.82×10^8 m
Distance to the Sun :	1.50×10^{11} m
Radius of the Earth :	6.37×10^6 m
Radius of the Moon :	1.74×10^6 m
Radius of the Sun :	6.96×10^8 m

Printed in the United States
By Bookmasters